Plants: A Very Short Introduction

VERY SHORT INTRODUCTIONS are for anyone wanting a stimulating and accessible way into a new subject. They are written by experts, and have been translated into more than 45 different languages.

The series began in 1995, and now covers a wide variety of topics in every discipline. The VSI library now contains over 500 volumes—a Very Short Introduction to everything from Psychology and Philosophy of Science to American History and Relativity—and continues to grow in every subject area.

Titles in the series include the following:

Timothy Walker

PLANTS

A Very Short Introduction

OXFORD
UNIVERSITY PRESS

OXFORD

UNIVERSITY PRESS

Great Clarendon Street, Oxford ox2 6DP

Oxford University Press is a department of the University of Oxford.
It furthers the University's objective of excellence in research, scholarship,
and education by publishing worldwide in

Oxford New York

Auckland Cape Town Dar es Salaam Hong Kong Karachi
Kuala Lumpur Madrid Melbourne Mexico City Nairobi
New Delhi Shanghai Taipei Toronto

With offices in

Argentina Austria Brazil Chile Czech Republic France Greece
Guatemala Hungary Italy Japan Poland Portugal Singapore
South Korea Switzerland Thailand Turkey Ukraine Vietnam

Oxford is a registered trade mark of Oxford University Press
in the UK and in certain other countries

Published in the United States
by Oxford University Press Inc., New York

British Library Cataloguing in Publication Data

Data available

Library of Congress Cataloging in Publication Data

Data available

Typeset by SPI Publisher Services, Pondicherry, India
Printed and bound by
CPI Group (UK) Ltd, Croydon, CR0 4YY

ISBN 978-0-19-958406-2

Contents

List of illustrations

Chapter 1
What is a plant?

Plants, like love, are easier to recognize than to define. At the entrance to many areas of outstanding natural beauty in England can be seen a sign that asks visitors to avoid 'damaging trees and plants'. It is fair to ask in what way is a tree not a plant. A plant is often defined simply as a green, immobile organism that is able to feed itself (autotrophic) using photosynthesis. This is a heuristic definition for plants that can be refined if some more characters are added. Sometimes plants are described as organisms with the following combination of features:

1) the possession of chlorophyll and the ability to photosynthesize sugar from water and carbon dioxide;
2) a rigid cell wall made of cellulose;
3) storage of energy as carbohydrate and often as starch;
4) unlimited growth from an area of dividing and differentiating tissue known as a meristem;
5) cells with a relatively large vacuole filled with watery sap.

So trees are clearly plants, and it is not difficult to think of other organisms that are unequivocally plants even though they lack one or more of these characteristics. For example, the orchid *Corallorhiza wisteriana* has the flowers of an orchid, produces tiny seeds typical of the family Orchidaceae, and has the vascular tissue that you find in the majority of land plants. However, what it does not have are green leaves, because this orchid is mycotrophic, meaning that it lives off fungi which themselves derive their energy from decaying material in the forest floor. It is able to do this because of a very intimate relationship with a fungus, a characteristic found to varying degrees throughout the orchid family. In a similar vein, *Lathraea clandestina*, which can be seen growing on the banks of the River Cherwell in Oxford, has flowers reminiscent of a foxglove, yet it too has neither shoots nor leaves. Its flowers emerge directly from the soil because this plant has roots that are able to infiltrate the roots of willow trees and divert the nutritious contents of their vascular tissue. Both of these plant species have lost the ability to photosynthesize, but they are still plants because they share many, many other features with those plants which do still photosynthesize.

The problem with the definitions above is that they are too limited, because they do not take into account some of the algae that live in water. In order to arrive at a sensible and unambiguous definition for plants, we need to consider how we classify biological organisms. Similar individuals are grouped together into a species. Similar species are then grouped into a genus. Similar genera are grouped together into a family; and similar families are grouped into an order; similar orders into a class; similar classes into phylum; and similar phyla into a kingdom. Each of the groups in this hierarchy can be referred to as a taxon, and the study of groups is known as taxonomy. Prior to the 19th century, taxonomists tried to create a *natural* classification that revealed the plan of the Creator. Since the 19th century, biologists have questioned whether species can change and evolve by retaining those changes and passing them on to their offspring.

A great deal of work is currently being carried out to build the 'tree of life' (or phylogeny) that shows how all living organisms are related to each other. This work received its kickstart in 1859 with the publication of Darwin's *On the Origin of Species*, and it is still ongoing. An evolutionary tree is the only illustration in

1. *Orobanche flava* is one of many parasitic plants that do not photosynthesize but which steal from other plants

The Origin, and Chapter 13 of the first edition is still an eloquent introduction to taxonomy. Darwin talks about the possibility of building a *natural* classification, but now natural means revealing the course of evolution and not the mind of God. Classifications now are based on what Darwin called *commonality of descent*. All the members of a taxon must share a common ancestor, and the group must contain all of the descendants of that ancestor. If these criteria are fulfilled, then the group is said to be monophyletic. Monophyletic groups occur at every rank in the classification from species to kingdom.

If we see the evolution of species over the past 3,800 million years as a branching tree, then plants are one set of the branches on the tree of life, and this set of branches is all connected back to one crutch. The arguments start when you try to decide which crutch marks the start of plants. It is worth saying at this point that fungi are definitely *not* plants. Fungi are in fact on the branch next to animals on the tree of life. Despite this, mycologists (who study fungi) do tend to be grouped with botanists rather than zoologists in university departments.

The original plants

At the heart of any definition of plants is the ability to photosynthesize. Unfortunately, there are organisms that photosynthesize but which cannot be considered by anyone to be plants. In particular, there are the photosynthetic cyanobacteria.

It is currently believed that life has evolved just once and that this happened about 3,800 million years ago. At that time, the world as an environment for biology was very different. There was no protective ozone layer to absorb the harmful ultraviolet light from the Sun. Furthermore, the atmosphere contained a great deal of carbon dioxide but very little oxygen.

The first living organisms were simple compared to the majority of plants that we see around us today. For a start, they were unicellular. They were prokaryotes. There are many prokaryotic organisms still extant in two big groups: the archaea and the bacteria. (The other major group of organisms are the eukaryotes, that is, plants, animals, fungi.) Fossilized prokaryotes have been found in rocks dated at nearly 3,500 million years old. The fossils of these early bacteria are grouped in structures that look the same as the stromatolites that can be seen in several places around the world today.

A stromatolite is a cushion-shaped rock that is found on the edges of warm shallow lakes, most commonly salt-water lakes, and they are (very simply) laminated accumulations of microbes. The colonies of the unicellular cyanobacteria live in a film of mucus. Calcium carbonate builds up on the mucus and the cyanobacteria migrate to the surface and a new layer of mucus is formed. These alternating layers are then fossilized and the bacteria enclosed in the rocks. So it was clear to see that prokaryotic life had evolved perhaps as early as 3,800 million years ago, but it was not so easy to determine how these early living entities found the energy to live. Some may have synthesized enzymes to break down minerals, but this was slow. There is now compelling evidence that the cyanobacteria in these fossil stromatolites were able to capture the energy of the Sun and use it to synthesize molecules containing carbon derived from the abundant carbon dioxide in the atmosphere. This evidence is based around the fact that the enzyme that drives the capture of carbon from carbon dioxide preferentially fixes one carbon isotope (^{12}C) over the other that is also present in the atmosphere (^{13}C). So if carbon compounds contain the two isotopes in different proportions from those in the atmosphere, then the compounds were the product of photosynthesis. Carbon compounds have been found in rocks in Greenland that have the carbon isotope ratio produced by photosynthesis.

Photosynthetic organisms, with which we are familiar, use water as a source of electrons. The oxygen in the water is then released into the atmosphere as gas. It is thought that the first photosynthetic cyanobacteria may have used hydrogen sulphide (H_2S) rather than water (H_2O). It is currently believed that by 2,200 million years ago, cyanobacteria were generating large amounts of oxygen and that this was accumulating in the atmosphere. This may seem like a small point, but the fact that cyanobacteria began using water as a supply of electrons led eventually to the levels of oxygen in the atmosphere that made aerobic respiration possible and the majority of biology as we know it. The generation of oxygen had another effect, namely the formation of the layer of ozone in the upper atmosphere, whose absence has already been noted and whose protective function is so important for biology. Prior to this, the mucus in the stromatolites may have helped to protect the cyanobacteria. Living in water would also have afforded some protection.

So to recap, we see that by 2,000 million years ago, there was a large population of prokaryotic cyanobacteria that was generating oxygen by photosynthesis, but there was still nothing that we could describe as a plant. The evolution of plants required an event that must have happened but for which we do not have a complete cast list. This event was the formation of the first eukaryotic cell. Eukaryotes cells are more organized internally than prokaryotes. They have organelles enclosed by membranes such as the nucleus and mitochondria and, in the case of plants, chloroplasts. Organelles are just small 'organs' found inside cells which perform specific functions within the cell.

It is believed that 2,700 million years ago, an unidentified unicellular prokaryotic organism engulfed another but did not break it down. The engulfed cell retained its membrane and gave up some, but not all, of its genes to be included in the nucleus of the host cell. This engulfing followed by 'cooperation' is known as endosymbiosis. This early eukaryotic organism (known as the proto-eukaryotic cell) lived by metabolizing photosynthetic products from free-living cyanobacteria. The evidence for this endosymbiosis is simple: the organelles have two membranes – their own and one from the host that engulfed them. The evidence for the timing of the first endosymbiosis is equally simple. One of the unique features of all eukaryotes is the production of sterols. When a eukaryote dies and breaks down, the sterols are converted into steranes, and these persist in rocks for a very long time. Rocks 2,700 million years old contain steranes, and so there is a trace of dead eukaryotes but no fossils of intact organisms.

Many years passed and the diversity of eukaryotic organisms increased, resulting in evolutionary lineages that produced many other species (both extant and extinct) but not plants. However, having recruited one type of prokaryotic organism, the proto-eukaryotic cell recruited another, and this time it was a photosynthetic cyanobacterium. As before, the incoming organism became an organelle, and some, but not all, of its genes were transferred to the nucleus of the host cell. Again, as before, the organelle, known now as a chloroplast, has a double membrane.

The oldest fossil evidence of the structure of a probable eukaryote is in rocks 2,100 million years old. The organism, named *Grypania spiralis*, has no extant descendants. It looks a bit like an algae, and so it is believed (or perhaps hoped!) that it was photosynthetic. At 2 millimetres in diameter, it is big enough to be an ancestor of some of today's algae, but it cannot be proved to be this ancestor. The first undisputed fossil of a photosynthetic eukaryote that can be placed in an extant taxon has been found in rocks 1,200 million years old. The organism, *Bangiomorpha pubescens*, is a red algae and is thus named because it resembles the extant red algae *Bangia atropurpurea*. In addition to looking similar, these two species also share a habitat, the margin of land and water.

Bangiomorpha is significant for another reason: it is currently the earliest example of a multicellular eukaryotic species that not only has cells with specific functions, but also one of the functions of these specialist cells is to indulge in sexual reproduction. Multicellularity is one of those important biological events that has evolved more than once on different branches of the tree of life. The most recent common ancestor of plants and animals was unicellular and yet both are now dominated by multicellular organisms.

The fossils of *Bangiomorpha* are so well formed that it has been possible to reconstruct its life cycle and it is similar to some of those found in the red algae. The spores germinate and grow into the multicellular body of the plant. The spores contain only one set of chromosomes (that is, they are haploid), so the algae plant is haploid. At the base of the plant is a holdfast that fixes the plant tightly to a rock. At the top, the plant becomes flattened, and as it grows upwards it is able to capture more light. Some of the cells in this thallus differentiated into haploid gametes that are a prerequisite for sexual reproduction.

So at some time between 2,100 and 1,200 million years ago, the first photosynthetic eukaryotic organisms emerged. These were the first plants, and every subsequent organism on this branch of the tree of life is a plant. The two endosymbiotic events that define this branch happened once and are known together as the primary endosymbiosis. The evidence for this has been derived from the analysis of DNA sequences. This technique, available from 1993, has been very important in unravelling previously tangled problems of evolution.

So plants as described in this book are monophyletic. Intriguingly, it is becoming clear that there has also been a secondary endosymbiosis in which some of the true plants have been incorporated into non-plant organisms and the resulting organisms are also not plants. Perhaps the most familiar of these

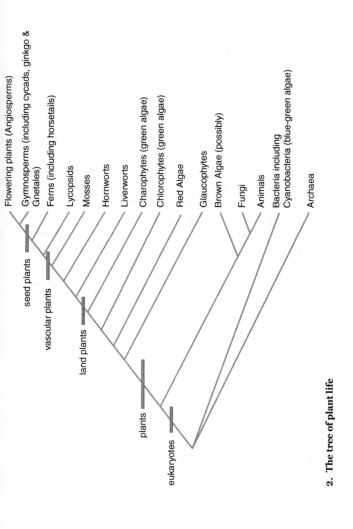

2. The tree of plant life

are the brown algae such as kelp. So if you visit a beach and start looking at the 'seaweeds' left behind by the tide, the green and red ones are plants and the brown ones are animals.

The oldest extant plants

The lowest side branch on the plant branch of the tree of life consists of a very old (at least 1,200 million years), very small (13 species) group of tiny (microscopic) freshwater algae in three genera – *Glaucocystis, Cyanophora, and Gloeochaete* – known collectively as the glaucophytes. The lowest branch on a phylogenic tree contains the organisms that are thought to have changed the least of any others on the tree. This does not mean that they look exactly like the original ancestors. The evidence that supports their basal position is two-fold. Firstly, in addition to the two membranes around the chloroplasts, there is a peptidoglycan layer. This is similar to the envelope found around bacteria, and the chloroplasts are sometimes referred to as cyanelles to distinguish them from those found in the rest of plants. This peptidoglycan layer is not found in any other plants, the inference being that it has been lost early in the evolution of plants. Secondly, they have pigments called phycobilins in their plastids. Plastids are organelles found in plants where important chemicals are manufactured or stored. These pigments are only found in cyanobacteria, the glaucophytes, and the red algae, which are the next side branch on the plant branch of the tree of life. In all three groups, these pigments are bound into phycobilisomes. In addition to the phycobilins, they have chlorophyll a.

This small group is of interest to evolutionary botanists because it is thought that they are the closest extant species to the original endosymbionts. Some species can move and some cannot, while some have cellulose cell walls and some do not. Sexual reproduction is unknown in this group of plants. The evidence from DNA sequences confirms the morphological evidence, and so clearly these little plants are strong contenders for the title of the earliest diverging branch on the plant tree of life. The next side branch up is the red algae. It is worth reiterating that the term 'algae' is a vernacular term that is applied to a group of organisms, some of which are plants (red algae and green algae) and some of which are not plants (blue-green and brown algae).

Red algae

There are many species on this branch: somewhere between 5,000 and 6,000, and perhaps as many as 10,000, of which just a small percentage live in fresh water. The red algae share some characteristics with the glaucophytes that have subsequently been lost and so do not appear in the rest of the plants. These characters are phycobilin pigments and phycobilisomes and having just chlorophyll a. These pigments, other than the chlorophyll, give these plants their distinctive red colour. Red algae have other features in common such as storing energy as glycogen (or floridean starch). Glycogen is a large molecule with lots of glucose in a chain off which come side chains of molecules. Some species secrete calcium carbonate and these are important in the construction of coral reefs, hence why these are known as coralline algae. Red algae all have a double cell wall. The outer layer is economically important because it can be made into agar, which has many uses including in cooking. The internal wall is made in part of cellulose, like most plants.

As one would expect, in a taxon as species-rich as the red algae, there is a lot of diversity, but there are common patterns to their life histories. However, that pattern is complicated and very different from the life cycle of mammals with which we are most familiar. This familiarity colours our preconceptions and assumptions about plant reproduction, and it can create a barrier that makes understanding plant life histories much more difficult than it need be.

The first false assumption is that *every* free living organism needs two complete sets of chromosomes just because we do. This is not true, and we have already seen a fossil species (*Bangiomorpha*) which spent much of its life in the haploid state with one of each chromosome, rather than in the diploid state in which it would have had two sets. It is reasonable to ask if there are any advantages or disadvantages attached to being haploid or diploid. A fact of life as a result of being haploid is that any deleterious mutation will be expressed and the organism may perish as a result. So it may appear that having two copies of each gene is better because the harm of a deleterious mutation can be overcome by the good copy, or perhaps the combination of two versions of a gene might be better than just one version. However, this is a twin-edged sword because it means that diploid cells can build up many potentially crippling mutations. Given that there appears to be an evolutionary trend towards diploidy in all the major taxa in the tree of life, it appears that this is, however, a more stable evolutionary strategy. Despite this, it cannot be denied that being haploid has not prevented many taxa from surviving successfully for hundreds of millions of years. Among these taxa are red algae, green algae, mosses, liverworts, and ferns.

The second false assumption that we make is that there will be a group of cells (the germ line) that from a very early stage in the life of an organism will be responsible for making the gametes – the sperm and eggs. This is true in mammals and many other animals but not in plants. In general, the development of plants and the differentiation of their cells into different structures is very flexible and certainly not determined early in the life of the embryo. This can be seen clearly when a gardener roots a cutting, or when a biennial plant like a foxglove changes from vegetative growth and starts to produce flowers. Plants do not have a group of germ cells. Instead, they have a distinct and finite stage in their life histories when a haploid plant produces sperm and/or eggs. Because the plant is already haploid, there is no need to halve the number of chromosomes before the gametes are formed by differentiation and mitosis (cell division whereby the chromosome number remains the same and two identical cells are produced). The haploid stage in the life history produces gametes, and so is very sensibly known as the gametophyte, '-phyte' being derived from the Ancient Greek word φυτόν, meaning plant.

A third false assumption (that is clearly wrong given the previous paragraph) is that diploid plants will produce haploid gametes by the process of meiosis (cell division whereby the number of chromosomes is halved). Instead, the diploid plants produce haploid spores by meiosis. Unsurprisingly, the diploid stage in the life history that produces spores is known as the sporophyte. These haploid spores grow into the haploid gametophyte, and the life history repeats itself. Plant life histories therefore consist of alternating stages or generations: between a haploid generation and a sporophyte generation. Plants are often described as demonstrating an alternation of generations.

Red algae (or their ancestors) were the first plants to display sexual reproduction, and so it is worth describing it here as it makes what comes later easier to understand. What follows is a description of the life history of *Polysiphonia lanosa*. This is a red algae that may be familiar to anyone who has spent time rock-pooling in the intertidal zone of the coast of Britain and Ireland. This species of red algae grows on the outside of *Ascophyllum nodosum* (a brown algae and so not a plant) which is very common all the way up the west coast of Europe and the north-east coast of North America. The *Polysiphonia* is probably an epiphyte on the *Ascophyllum*, but some people have recorded that the former actually penetrates the latter, and this would make it parasitic. The *Polysiphonia* plants look like cheerleaders' pompoms on the surface of the *Ascophyllum*.

These plants of *Polysiphonia* may be male or female gametophytes, but they look very similar. The male plants

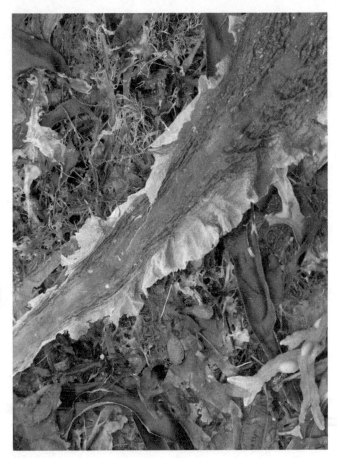

3. **Brown, red, and green algae are often washed up on the seashore together, but only the red and green algae are plants**

produce male gametes. These are known as spermatia rather than sperm because they do not have a flagellum, or tail, for propulsion. These are released into the water around the plants and the hope is that they will find a female gametophyte. The female gametophyte produces (but does not release) a female

14

gamete. This is a cell that is retained inside a structure known as the carpogonium which consists of the female gamete and a trichogyne. This trichogyne is a protuberance whose function is to catch a passing spermatium. Once caught, the spermatium will donate its nucleus to that of the female gamete, and a diploid cell is formed known as a zygote.

The zygote develops into a diploid structure that remains enclosed in, and therefore completely parasitic on, the haploid female gametophyte. The diploid structure is known as the carposporophyte, from which we can deduce that this pustule produces diploid carpospores. These carpospores are released into the water and drift around hoping to land on a suitable substrate, which in the case of *Polysiphonia* is an *Ascophyllum* frond. These diploid carpospores germinate and grow into diploid tetrasporophytes. Rather unhelpfully, these tetrasporophytes look very similar to the gametophytes despite the former being diploid and the latter haploid. When a species has haploid and diploid generations that look similar, it is said to be isomorphic. When the two generations look different, they are said to be heteromorphic. Some red algae are isomorphic and others are heteromorphic.

When mature, tetrasporangia form on the surface of the branches of the tetrasporophyte. The tetrasporangia are the sites of the production of haploid tetraspores, so called because they are formed in tetrads (or 4s). This means that they are joined together in a triangular pyramid with each spore attached to each of the other three. These haploid spores are produced by a diploid plant and therefore they are the result of meiotic cell division, whereby the number of chromosomes is halved. The tetrad of spores is released, and hopefully it will land on a frond of *Ascophyllum*. One life history of *Polysiphonia lanosa* is now complete.

The internal structures of red algae are as variable as their life cycles, but *Polysiphonia lanosa* can be used as an example. Each branch has a central axis of elongated can-shaped cells that are joined end to end. These cells are joined by pit connections that form during the process of cell division. Associated with these connections are pit plugs, which are able to seal the connection should one of the cells die. Intercellular connections are an important component of multicellularity. Around this central axis of cells is a layer of periaxial, or pericentral, cells. These are the same length as each cell in the central core and aligned with the cells therein, thus making the branches of the plants look like a series of repeating units. There may be a further layer of cells, known as the cortical cells, around the periaxial layer.

Green algae

Green algae belong to a much larger and very diverse group of plants called simply the green plants. It is currently believed that they have all evolved from one common ancestor. They share a number of features, including both forms of chlorophyll – a and b – and they have a cell wall made of cellulose. The majority of the green plants are the land plants; the rest are the green algae.

In the same way that 'algae' is a vernacular term that is applied to a group of organisms, some of which are plants (red algae and green algae) and some of which are not plants (brown algae), the term 'green algae' refers to two different groups that do not share one unique common ancestor. The green algae are two branches on the evolutionary tree: the chlorophyte algae and the charophyte algae. The evolution of these plants and their relationships is not yet fully understood; in fact, it is a bit of a mystery because the land plants have received much more attention than the algae in the past twenty years. However, it appears that one branch is very much less species-rich than the other: the charophytes, including the Charales. This smaller group is arguably more important in that it is the sister group of the land plants that now dominate the world.

Rather than try to give a comprehensive survey of the green algae, a few species will be described in detail to illustrate the diversity found in this group. Green algae are found in fresh water *and* sea water. Many species are unicellular but some are filamentous, while others spend some of their time as single-celled organisms that then form a multicellular colony in which some cell differentiation occurs. *Volvox carteri* is a good example of this latter behaviour. Some algae form symbiotic relationships with fungi known as lichens. While the algae can live without the fungus, the reverse is not true, and so lichens are known by the fungus's name not the algae's. Furthermore, one species of algae can form a symbiosis with many different fungi. (Lichens are *not* plants.)

The first group of green algae many people encountered was *Chlamydomonas* because it grew in the school pond. This unicellular genus, along with the *Volvox* and *Ulva* described later, belong to the chlorophytes. *Chlamydomonas* has two flagella and has been used extensively as a 'model organism' in the movement of flagella. The adult plant is haploid, with just one set of chromosomes. This cell can reproduce asexually by simply dividing mitotically. Before these new adults are produced, the *Chlamydomonas* lose their flagella and group together. The cells then divide but in an uncoordinated fashion, producing new unicellular organisms.

However, the adult *Chlamydomonas* can differentiate into a gamete. In some species, the male and female gametes are the same size (isogamous), whereas in others the females are larger (oogamous). The two gametes fuse in the water and form a diploid zygote that is encased in a thicker wall that can protect the young zygote. This zygospore can withstand harsh conditions, but when it is feeling that conditions are correct, it will divide meiotically and four new adult plants are released from each zygospore and the life history is completed.

Chlamydomonas is in the family Chlamydomonaceae which is in the same group of families (or order) as the next genus we shall look at – *Volvox*. *Volvox*, like *Chlamydomonas*, has been investigated in depth because it is capable of existing as a unicellular plant or in a spherical colony of several thousand cells. *Volvox carteri* is thus a useful subject if you want to study the evolution of multicellularity, which has evolved many times in plants, let alone independently in plants and animals.

So the unicellular *Volvox* plants come together to form a colony whereby several thousand individuals become embedded around the outside of a gelatinous ball of glycoprotein (known as a coenobium) that is up to 1 millimetre in diameter. This colony works as a collective, with coordinated flagella beatings that can move the ball towards the light. Sometimes connecting strands of cytoplasm can be seen connecting the cells. The cells perceive the light through eyespots which are more common on one side of the colony than on the other, thereby giving the ball a front and back, or anterior and posterior pole. This coordinated group of individuals then becomes a truly multicellular organism when some cells begin to divide asymmetrically to give one small and one larger cell. The two new types of cell are incapable of an independent existence. The small cells are somatic cells whose function is to propel the colony with their flagella. The larger cells, known as gonidia, accumulate at the posterior pole, where they divide and give rise to daughter colonies. These young colonies are initially retained inside the coenobium, with their flagella orientated inwards, but when the parent colony ruptures to release the offspring, the cells reorientate and the flagella are on the outside of the sphere. Vegetative reproduction is complete. Bizarrely, sometimes granddaughter colonies form inside the daughter colonies before they are released by the mother colony.

Sexual reproduction of *Volvox* is also different from that of *Chlamydomonas*. Some species have colonies that produce both sperm and eggs (monoicous), while in other species the colony will only produce sperm or eggs (dioicous). (It should be noted that this is different from monoecious and dioecious which is used when diploid plants are male or female as opposed to bisexual.) At the commencement of sexual reproduction, some of the generative cells in the colony will *either* begin to produce sperm which are released *or* to develop into egg cells which are not. The sperm are produced in sperm packets, which are simply bags of sperm that are released from the parental colony. There is some evidence that these packets release a pheromone to make other colonies sexually active. When the sperm finds an egg and successfully fertilizes it, a thick-walled spore results that contains the zygote. This diploid spore, known as the meiospore, is capable of withstanding harsh conditions, but in the correct circumstances will undergo meiosis and release haploid offspring.

One of the more familiar green algae to anyone who has been rock-pooling along the shores of temperate regions of the world is sea lettuce, or *Ulva lactuca*. This often scruffy-looking plant is found attached to rocks by a round holdfast or drifting freely. The rest of the plant is up to the size of a dinner plate and is a thallus of just two layers of cells thick, making it very flimsy, and so it regularly gets torn by the action of the waves on the rocks. However, this is not a problem as it lives in water and the plant is buoyant and supported by the water. The cells in each layer are arranged randomly and any one of them can divide. This means that there is no equivalent of the meristems that we find in flowering plants. The individual cells are not interconnected in any way, making this little more than a colony in some people's minds. This is an over-simplification as the plants are in fact more organized than that, in that they are like herbaceous perennial land plants because they can regrow from the holdfast each spring or if the thallus breaks off. The pieces of thallus that break off have been found to form new plants in laboratory conditions, but it is not known if this happens in the wild.

4. Sea lettuce is one of the most common green algae on English beaches

Ulva, like all sexually active plants, has a life history that includes an alternation between a haploid generation and a diploid generation. The twist in this part of the tree of life is that the haploid and diploid generations look the same and so are described as isomorphic. The haploid plants produce gametes. To do this, cells around the margin of the gametophyte thallus divide and differentiate into biflagellated sperm or eggs. These are similar morphologically except that the females are slightly larger. Both male and female gametes are able to photosynthesize and to swim towards a light source (positively phototaxic). This means that the gametes swim up to the surface of the water. It is believed that the flagella are more than just a means of propulsion; they are implicated not only in sexual identity but also as facilitators of adhesion once a gamete of the other sex has been located. The gametes of both sexes have eyespots at the base of their flagella. The gametes differ in that the female eggs have 5,500 particles in the outer membrane of the eyespot chloroplast while the male gametes have 4,900. Having found the gamete of their dreams, the now quadriflagellate zygote is negatively phototaxic, meaning that it swims away from the water's surface to the rocks at the bottom of the water, where they can grow a holdfast and then a new thallus, only this thallus and holdfast are diploid.

When mature, the diploid thallus of the sporophyte produces haploid spores from the margin. These are the result of meiotic division that halves the chromosome number. These zoospores (like the zygotes) are both quadriflagellate and negatively phototaxic. Also like the zygote, the eyespot membrane has 11,300 particles, and this is thought by some to have a role to play in phototaxis. The zoospore will then grow into a holdfast and a male or female gametophyte. *Ulva* is collected from beaches in Scotland and eaten in soups and salads, and in Japan it is used in some sushi dishes.

The charophytes are a much smaller group than the chlorophytes in terms of the number of species. One may be familiar from school days. *Spirogyra* is a filamentous algae found in freshwater pools and ponds. It normally lives below water, but in warmer weather the rapid growth rate and lots of oxygenic photosynthesis results in a frothy, slimy mass of tangled filaments rising out of the water. At any time of the year, it is easily identified by the fact that the chloroplasts are arranged in a pattern that resembles a stretched spring. The cylindrical cells are joined end to end and individual filaments may be many centimetres long. The cell wall has two layers: an outer coat of cellulose and an inner wall of pectin. The filaments can break, but this is essentially asexual reproduction as the two halves can each grow into a new plant.

Each cell of an adult plant of *Spirogyra* is haploid, so it is a gametophyte. Sexual reproduction is simple and comes in two scenarios. In the first, two different filaments come to lie alongside each other. Tubes grow out from cells in each filament and fuse at their tips to create a conjugation canal between two cells, one from each filament. The contents of the male cell migrates into the female cell, the nuclei fuse, and a diploid zygote is formed and released as a zoospore. This is known as scalariform conjugation, like a ladder. The other scenario is known as lateral conjugation. This is when a filament forms conjugation tubes between adjacent cells in the same filament. This is followed by the migration of the male contents into the female cell and the formation and release of the diploid zoospore as happens in scalariform conjugation. This spore then divides meiotically to give four haploid cells, from which new gametophyte filaments form.

The final example for this chapter and the second charophyte is *Chara* itself. This is a multicellular plant that is found in freshwater pools in temperate regions of the northern hemisphere. The plants look similar in general terms to other water plants such as *Ceratophyllum* (see Chapter 5) and to some land plants such as horsetails or goosegrass, yet they are closely related to neither. The plant consists of a central filamentous stem from which whorls of branches are produced at regularly spaced nodes. The plants may be found floating freely, but they do grow into the mud at the bottom of ponds with the production of rhizomes. The cells at the apex of the stem divide, with the upper daughter cell retaining the function of apical cell. The apical cell in a plant is simply the cell at a tip. The lower daughter cell develops into either a nodal or inter-nodal cell. It is from the nodal cell that the whorl of branches grows. The branches are either short and of determinate growth, or long and of indeterminate growth.

This plant is the haploid gametophyte. The plant produces motile sperm that are released into the water. The female gametes are not released but retained in structures on the gametophyte. The diploid zygote that is formed could, and perhaps through our eyes should, grow into a diploid sporophyte, as happens in *Ulva*. However, *Chara* is not us (that is, not human) and the zygote goes straight into meiosis to produce four haploid spores. These spores drift away and develop first into a filamentous protonema and then into more haploid adult gametophytes.

So the green algae consist of two separate groups: the chlorophytes and the charophytes. These are not one monophyletic group sharing one unique ancestor; rather, they are adjacent branches on this limb of the tree of life. There is one more big group left on this limb and it shares a unique common ancestor with the charophytes. This big group is very big. It consists of about 400,000 species. This group of plants is very familiar to us because they are the land plants. These plants can be traced back to one unique, unidentified ancestor that had accumulated a new combination of traits that enabled it to survive for most of the time out of water. This ancestor was the first land plant, and without it there would be no terrestrial ecosystems as we know them, and there would certainly be no *Homo sapiens* and no Oxford University Press and no *Very Short Introduction* to plants. The next chapter is about the land plants.

Chapter 2
Living on dry land

About 470 million years ago, a plant survived for more than one generation out of the sea water where the green algae had lived for many years. This is one of the major events in the natural history of the Earth and is often referred to as the invasion of the land. However, it could be just as easily (and perhaps more accurately) described as the invasion of the air, since many of the plants in the water had already evolved means of attaching themselves to the bed rock. So what were the challenges facing plants wanting to leave the familiar comfort of the sea?

Firstly, there was desiccation resulting from the wind experienced when living in the air. When living in the sea, there was water all around that would enter the plants' cells by osmosis. It could be said that keeping water out is more of a problem than keeping it in if you live in the sea. When the first land plants were stranded out of the water, they had not only to reduce loss of water but also to find a means for taking up water to replace the inevitable loss. So hand in hand with the problem of taking up water, there is the problem of holding onto the water. If we look at how some of the older lineages of land plants, such as mosses, have solved this problem, we find that rather than preventing desiccation, they simply tolerate it and then rehydrate when it rains. It should come as no surprise that 10% of the moss species found in Europe are found in Britain, and half of European liverwort species are found in Ireland, because both countries are notorious for rainy weather. So the first land plants possibly led a rather Jekyll and Hyde lifestyle, absorbing water and nutrients when it was wet and shutting down when it was dry.

Secondly, and connected to the first problem, the first land plants had to find a way to allow air (containing carbon dioxide and oxygen) into their photosynthetic structures. For plants living in water, these two substances are dissolved in the water and sufficiently available. Any openings in the leaves for the ingress of gases will also be an opening for the exit of water, which is now very precious.

Thirdly, and related to the previous problems, there was the problem of finding the other raw materials that a plant requires such as nitrogen, phosphorous, potassium, calcium, iron, sulphur, magnesium, silicon, chlorine, boron, zinc, copper, sodium, molybdenum, and selenium. It must be appreciated that the land these plants were colonizing was bare rock. While there may have been deposits of sand and other fine particles as well as larger pieces of rock, there was probably no soil as we now know it. This is because a vital component of today's soils is the organic matter that is the remains of dead organisms. There are also the other living organisms themselves such as bacteria, fungi, and animals. Before there was soil, there was no evolutionary pressure to develop a complex root system to take up nutrients from the soil. There was, however, the need for an anchor that would prevent the plant from being blown or washed from its place on the rocks. The algae have a holdfast but this structure does not seem to have made its way onto dry land.

Fourthly, there was the problem of finding new places to live. Although the plant wants to be attached securely, it still wants to be able to explore the area for potentially better positions and to escape from disasters. This required the emergence of protection for the haploid spores. It turns out that this may have already been in place in the form of sporopollenin. This is a very complex and extremely durable substance found on the outside of the spores (and pollen) of land plants and in the walls of spores of a few green algae such as *Chlorella*. For people studying sporopollenin, its durability is a problem as it makes it very unwilling to reveal its structure and chemical composition, so it is impossible, at present, to be sure that the green algae sporopollenin is exactly the same as that found in land plants. It appears, therefore, that there was a pre-adaptation to solve this problem. Pre-adaptations are not uncommon and are probably a requirement for large advances or multiple innovations in evolution that would otherwise require the simultaneous emergence of several novel structures.

In this instance, one person's problem turns out to be a gift for other scientists, as the durability of sporopollenin enables evolutionary botanists and ecologists to track the distribution of plants by the deposits of pollen in soil cores and even ancient rocks. Some of the earliest evidence for the existence of plants on the land is spores showing liverwort characteristics that have been found in rocks dated at 475 million years old. These spores are 50 million years older than the first fossils of intact plants (or even fragments of plants).

The fifth problem is the delicate matter of the sperm finding an egg. The plants that live in water release one gamete (or occasionally both) into the surrounding water and it (or they) swims off to find a friend. For some botanists, this is seen as a very big problem for land plants, but this may be an exaggeration. The argument suggests that if you need water for your sperm to swim through on its way to the egg of its dreams, then the parent plants are going to be restricted to wet places. There are still some plants, for example mosses, for which water is required to facilitate the bringing together of gametes. Simple observations of where these plants grow show that they are simply restricted to places that are *seasonally* wet. Mosses are often found on the tops of walls, a habitat that may resemble the bare rocks on which the first land plants grew. While the tops of walls are dry in the summer, they are not in the winter, and so this is when mosses' gametes get together. Furthermore, by suggesting that mosses are weak because they need water for their sperm to swim through implies that other plants do not need water, and to suggest that any plants do not need water is a nonsense. Mosses have not found the temporary need for water to be a critical failing; they have been living on land for more than 400 million years.

If these five problems are put together, it is possible to see that one way of solving them simply and simultaneously is to be a liverwort. Gardeners are familiar with these little plants as one species, *Marchantia polymorpha*, that frequently grows on the top of pots in which one has sown seeds. These plants resemble a green liver-like sheet or thallus (though some other species of liverworts have ranks of leaf-like structures along the sides of a simple vein). Like the mosses, the liverworts are tolerant of desiccation. In the upper surface of the thallus, there are barrel-shaped air channels to supply the upper half of the thallus with the carbon dioxide necessary for photosynthesis and the oxygen for respiration. The lower part of the thallus is used for storage of substances such as starch. *Marchantia polymorpha* has unicellular rhizoids that can grow into the tiny fissures in the rocks to hold it in place. These rhizoids do not need to take up water and nutrients because these can be absorbed directly by the plant. As described already, the spores are coated in decay-resistant sporopollenin, and the plants restrict their sexual activity to the wet periods of the year to get round the perceived problem of releasing sperm straight into the environment. One feature of all mosses and liverworts is that they do not grow tall when compared to most ferns, conifers, and flowering plants, and in fact also when compared to many of the green and red algae that preceded them. This is because there is one more problem facing land plants that the seaweeds did not have – gravity. The support provided by the sea to buoyant algae has enabled them to grow into large structures. The air does not

provide this support, and the mosses and liverworts therefore
stay small, but again it has not prevented them from surviving
for hundreds of millions of years. The mosses evolved a very
simple internal pipework consisting of hydroids that are
sometimes found in fossil plants such as the *Cooksonia*
described below.

The question arises, how did plants like *Chara*, or an ancestor of
modern-day *Chara*, evolve into *Marchantia* or something like it?
We might try to find evidence of an amphibious plant that has a
free-floating aquatic form but which can live on mud if the pond
in which it lives dries up. Fortunately, there is one such liverwort,
Riccia fluitans. This could be regarded as a half-way land plant, or
missing link, possessing some of the adaptations for living on land
in the air but not all in one go. This is, however, pure speculation
and not supported by fossil evidence. Sadly, the fossil evidence for
the invasion of land and air by plants is generally poor because
these soft little plants did not make good fossils, though, as said
before, the spores do make good fossils, albeit very small ones.

So the first land plants were liverwort-like and had thrown off the
dependence on water as an omnipresent, all-embracing growing
medium. This has to be regarded as one of the major landmarks in
the evolution of life on Earth because not only did these plants
radiate out and evolve into the 400,000 species of land plant that
we now see around us, but also they in turn provided many novel
habitats for animals, bacteria, and fungi. As ever, evolution did
not rest on its laurels but from the liverworts' ancestors evolved
many other groups of plants: some extinct, some extant.

Stems

As plants have evolved, different groups have taken ecological
centre stage for differing lengths of time. For many millions of
years, the plants were no more than a few centimetres tall. The
reason for this was two-fold. Firstly, the plants needed a rigid

internal scaffolding to keep them upright against the oppression of gravity. Secondly, they needed pipework through which water and other nutrients could move to ever-increasing heights. In addition, a pump to push the water up the plumbing would help. If you want to gain height, a cell wall and water pressure can get you only so far – a few centimetres tall at best. There is good fossil evidence that 425 million years ago, there were at least seven species of *Cooksonia* (now extinct) around the world which did have stems that varied in diameter from as little as 0.03 to 3 millimetres.

The fossils show that these plants consisted of dichotomously branching stems that terminated with a sporangium. This means that these were the sporophyte part of the life history. (Sadly, there are no fossils of anything that can be considered the gametophyte.) On the surface of these branching stems can be seen stomata for the ingress of gases and the loss of water. The loss of water is not all bad if water is plentiful and if its evaporation can be used to pull water up a narrow capillary-sized tube. In the fossils of *Cooksonia*, there appear to be tubes through which water could be pulled in this way. These are known as xylem tubes, though they may be more like the hydroids found in extant mosses than the xylem in extant vascular plants. It is assumed that the loss of water through the stomata was enough of a pull for water to move up this stem to the top of the plant where it was needed.

However, if plants were to get bigger, they had to come up with something new, or as often happens, have a look in the tool box and see if there was anything already out there in their genetic inheritance that was fit for purpose. What they found was lignin, which was first used by some of the red algae but then not used again. Just what red algae needed lignin for is anyone's guess. This is anthropomorphizing the evolutionary process whereby lignin was included in trees, *but* it is an illustration of one of the major strategies employed by plants, and that is that in order to survive they have to make use of what they have because they do not have the option of running away.

In order to understand the importance of lignin, we need to retrace our steps to the definition of a plant. One of the defining features is the possession of a wall around the cell. This has a number of functions such as restricting the volume of the cell to prevent explosion and preventing the ingress of some molecules. It is not very rigid in its basic form. The primary plant cell wall consists of two layers: an outer layer known as the middle lamella and the primary cell wall itself. The middle lamella is made up of a complex mixture of polysaccharides known as pectin. One of the functions of the middle lamella is to hold cells together in multicellular plants. It is also the part of the plant that breaks its hold when fruit and leaves are shed. Furthermore, it is very important in jam-making. The primary cell wall is made of a mixture of cellulose and hemicellulose. The former is perhaps the most prevalent non-fossil organic carbon molecule on Earth, accounting for one-third of all plant material. Cellulose consists of very long unbranched strands of glucose molecules. The hemicellulose, in comparison, is made up of shorter molecules that branch, and it is made up of more than just glucose. This primary cell wall is strong yet flexible, and when the cell inside it is at full turgor, the whole structure is quite rigid. The flexibility is important because as cells grow the primary cell wall has to grow too. The orientation of the cellulose microfibrils in the cell wall determines the direction in which the cell will grow. However, as we see when a plant wilts, if the turgor of the cell falls due to lack of water, then the plant becomes quite limp and the cell wall bends and folds.

A plant that wilted every time water was in short supply was never going to get taller except in places where it never stopped raining. This was where lignin was brought back in to make up the secondary cell wall. Lignin is strange stuff. It is second only to cellulose in its abundance; perhaps as much as 30% of the dry weight of wood is lignin, and 30% of all non-fossil organic carbon is lignin (until all the forests have been cut down). Chemically, lignin is odd as it lacks a consistent structure. However, this does not matter because it does the job. The job that it is required to perform differs from tissue to tissue and so does its structure. In many parts of the plant, it acts as a cement, filling the gaps between the molecules of cellulose, hemicellulose, and pectin. Lignin makes the cell wall rigid and hard. However, it can also be used as a storage facility such as in seeds, or as waterproofing when it contains a lot of suberin. This has an added advantage and this is that, when compared with the other components of the cell wall, lignin is hydrophobic, making a pipe lined with lignin a much more effective pipe than a pipe lined with cellulose, which is leaky.

Xylem tubes are made up of cells that have lost their living components and are empty. Xylem cells come in two basic designs. The first to appear (and therefore assumed to be the more primitive type) are tracheids. These were followed by the vessel elements. Tracheids are longer and thinner than vessels. They have a lignified cell wall that has a helical reinforcing layer. The tracheids are often associated with parenchyma, sometimes referred to as ground tissue. Where the tracheids join together, they have long wedge-shaped ends where the two flat sides rest against each other. This helps to prevent the continuity of the water stream from breaking as it rises up the stem. Tracheids tend to function in bundles with water leaking sideways from one to another, helping to prevent the inclusion of air bubbles in the stream of water.

The vessel elements are shorter and wider than the tracheids and more expensive to build. They line up very closely and the end walls that connect with the cells above and below them in the stem are perforated, making the passage of the water much easier. There is a variety of perforation patterns, but there is a penalty to pay for the increased speed of moving water and that is that the water stream can break more easily. Vessels were first identified in flowering plants and it was assumed for a long time that this was one of the reasons for the current success of the flowering plants. However, vessels have now been found in a range of plants including ferns, horsetails, and club-mosses, and so, like tracheids, they have evolved several times. Tracheids can hardly be described as a poor relation to the vessels when you consider that the tallest and largest trees have tracheids. The coastal redwoods at 360 feet tall are thought to be right at the limit of current plant engineering performance. It should be noted that the growth habit that we call a tree has evolved many times. A tree is very difficult to define precisely. In law, a tree is a single-stemmed woody plant that is at least 3 inches in diameter, 5 feet above ground level. Botanically, a tree is simply a plant with a stick up the middle – but how that stick is made varies.

Inevitably, with the evolution of these hard and heavy lignified stems came a number of other problems, including not only how do you stand up and also take up water and nutrients fast enough to supply the top of the plant, but just as importantly how do you get the products of photosynthesis down from the top of the plant to the roots where energy is required? Better roots were required, and this is discussed later, but some more plumbing was needed to transport the photosynthates.

When land plants were all mosses and liverworts, there wasn't a great deal of cellular specialization. Most of the cells were photosynthetic and so could look after themselves. Those that were not able to photosynthesize were so close to those that could that simple intercellular transport channels were sufficient. As soon as plants became lignified, the distance between the aerial parts and the subterranean bits was too long for diffusion to work. If the xylem is the ascending transport system, a descending transport system was required, and this is provided by the phloem.

There is one simple difference between phloem and xylem, and this is that phloem cells are still alive. Phloem as a tissue consists of three types of cells. Firstly, there are sieve-tubes, cells through which the sap flows. The cells have no nucleus but they do have a cytoplasm with a few organelles to enable them to fulfil their role. The sieve-tubes line up like sections of pipe. The end walls are like sieves, and through the holes pass larger than average plasmodesmata. Plasmodesmata are small tubes that connect cells to each other, and here they are found connecting the small amount of cytoplasm that there is to the companion cell, the second type of cell in phloem. This companion cell has an above-average number of mitochondria because it is having to power the sieve-cells. There is a specialized type of companion cell known as a transfer cell that has a much-convoluted cell wall, enabling it to gather solutes efficiently from the space in the cell walls around it. The final type of cell found in phloem is the ground tissue in the form of parenchyma, sclereids, and fibres. The sclereids are tough cells that are particularly common in plants in Mediterranean-type environments where the plants experience severe water stress in the summer. The cells in the sclerenchyma are nearly all secondary cell wall with a small amount of cytoplasm at the centre.

Many plants are short-lived, and once they have produced some progeny, they die. Others produce offspring year after year. Some of these perennial plants become woody and live for many years. This requires a stem that not only is resistant to pests and diseases but which is able to increase in girth. This scaffolding does not have to be living, and in many trees it is not. When a tree is felled, it is normally possible to measure the age of the plant by counting the number of concentric rings in the tree trunk. These rings of tissue are produced from a ring of vascular cambium that first develops behind the shoot tip. Cambium is a type of meristematic tissue that develops into other mature types of tissue. This tube of vascular cambium gives rise to xylem in the inside and phloem on the outer side. It also divides sideways to keep pace with the increasing girth of the stem.

In some trees, this does not work because the vascular tissue is not in a continuous ring around the stem, but it is arranged in bundles of xylem and phloem that are scattered randomly through the stem. This is the case in trees such as palms. Here, there is no secondary lateral meristem in the shoot as described in the previous paragraph. There is an apical meristem in the growing tip behind which there are leaf primordia, the meristems from which leaves grow. However, there is an extra structure called the primary thickening meristem. This is small in the shoot emerging from a newly germinated palm seed, but as the stem gets wider, then so does the primary thickening meristem. Its function is to produce vascular bundles, reminiscent of spaghetti coming out of a pasta machine. If the young palm tree is living in optimal conditions, the growing tip will get wider and wider until it reaches the maximum diameter for that species. The plant will then start to grow up and produce a proper trunk. However, if that plant suffers a period of less than optimum conditions in the future, the width of the meristems will decrease, resulting in a narrower stem. However, if things pick up again, the meristems can enlarge back to their former size, with the result that you get a tree with a waist.

When a seed germinates, the first structure to appear is usually the young root because the uptake of water is the first requirement of the seedling. If the seed is thought of as an embryo wrapped in a tough coat with a packed lunch in its pocket, then growing a root first is a good strategy. However, the packed lunch will only support the embryo for a finite period, and there is no going home to mummy for this embryo. It must produce its own food, and to do this it needs to photosynthesize, and to do this it need leaves (or something like leaves).

Leaves

The arrangement of tissues at the shoot tip is completely different from that found at the root tip. For example, there is no shoot cap because pushing through air should not result in damage. That is not to say that the growing point is not protected in many plants. A simple way of doing this is to have the meristematic tissue surrounded by the young leaves and their stalks. The shoot meristem gives rise to lateral appendages such as leaves as well as more stem, that grows into branches, and terminal organs such as flowers. This is different from roots, as the root does not generally produce any specialized structures just more roots. The control of the development of cells in a root is derived from the position of the cells in the root, whereas in the stem, more sophisticated and complex gene activity controls the operation.

The structure of stems and their cells has already been described, but a stem is basically a support system. Although green stems will contribute to the photosynthetic activity of the plant, it is rarely enough. Only in the case of cacti and some other succulent plants are photosynthetic stems all the plant needs because leaves are too wasteful to be a viable strategy when water is in short supply. However, in the majority of plants, leaves are produced. The arrangement of these leaves is normally constant within a species and is determined by interactions at the growing tip. The shoot meristem is in a constant state of fragmentation and replenishment. As a piece becomes detached from the side, it becomes a leaf primordia that will grow into a leaf with the bud that is found the base of that leaf. The next leaf will be formed when another piece of meristem becomes detached. Where this happens depends on how strong the inhibition is from the previous primordia. This point of minimum inhibition will be at different numbers of degrees around the stem depending on the leaf arrangement of the species in question. In plants with alternate leaves, it will be 120°; if the leaves are in opposite pairs, it will be 180°; and if spirally arranged, it will be approximately 137°.

We probably all think that we know what a leaf is. It is a flat green structure. We might remember from school that it has a thick cuticle on the cells of its upper surface, a thinner cuticle on the cells of the lower surface which is perforated by stomata, and in the middle is an upper layer of cells arranged like bricks (palisade mesophyll), below which there is a more open layer of cells (the spongy mesophyll). Running through all this are veins consisting of xylem, phloem, and probably some lignified stiffening tissue. This is a good generalized leaf structure, but here, as in many aspects of plant biology, there is no true default setting as plants have adapted to so many different ecological niches.

In actual fact, a leaf is a lateral structure on a stem, and in the axil where the leaf comes off the stem there is a bud that may grow into either a vegetative branch or into an inflorescence. The leaf is a determinate structure with a predetermined, finite size. Some 'leaves' are in fact spines (e.g. in some members of the bean family), some leaves develop into pitchers for catching dumb insects (e.g. *Heliamphora*), while other leaves are tendrils for aiding climbing. To be strictly correct, it is the other way around and spines, pitchers, and tendrils may be leaves. This is a simple

5. The leaves of *Heliamphora* are modified for the capture of insects

illustration of the principle of homology whereby function does not define a structure, rather identity of a structure is determined by its relative position and developmental origin.

In the same way that there are some leaves that do not look like leaves, there are other structures that look like leaves but are not leaves. The British plant butchers' broom (*Ruscus aculeatus*) has flowers and subsequently fruit growing from the upper surfaces of its leaves. Likewise, a Madeira climber, *Semele*, has flowers growing from the edges of its leaves. When one sees this type of thing going on, one needs to be very careful because all that is flat and green is not a leaf, and flowers cannot grow from leaves. These flowering leaves are simply flattened stems that superficially resemble leaves.

Branches

The stems of mosses and ferns usually branch dichotomously. This is relatively easy to achieve since it just requires the apical meristematic area to split into two halves. In most seed plants, the shoot meristem stays very much the same size during the growth of the plant, and lateral pieces of meristem become detached to form leaves with the associated buds that in their turn contain a lateral meristem. In some plants, the leaf primordia and lateral meristems leave the shoot meristem together but as distinct structures. In other plants, just the leaf primordia separates from the shoot meristem. It is later as the leaf begins to grow and differentiate that some of the cells revert to being meristematic.

Roots

It has already been noted that the little liverworts and mosses had little unicellular rhizoids, tiny root-like structures consisting of a single cell, to attach them to the rocks. Rhizoids are found elsewhere in the plant kingdom, including in the Characeae, so again this may be a pre-adaptation to live on land. It has recently been shown that the control of the development of rhizoids in the

6. *Ruscus aculeatus*, or butchers' broom, appears to have flowers growing on its leaves. These 'leaves' are actually flattened stems

moss *Physcomitrella patens* is controlled by the same genes that control the development of root hairs on flowering plants.

Tiny rhizoids alone were never going to have sufficient capacity on their own to supply water and nutrients to support a large perennial fern such as bracken, let alone a cactus, redwood, or oak tree. The biology of roots is a much neglected subject; out of sight really is out of mind in this case. Just as the parts of plants above the ground vary from species to species, so do the parts of plants below the ground. One thing that most of them (80% to 95% of all species) have in common is a close relationship with a bacteria or fungus. These arbuscular mycorrhizae benefit the plant by supplying nutrients, in particular phosphates. In return, the fungus takes sugars from some of the cells in the root.

This type of relationship appears in the fossil record and seems to have evolved on the underground parts of plants that we do not consider to have true roots. It is now thought that roots have evolved at least twice, not including rhizoids. The first roots are found in fossils 350 million years old. The plants to which these roots belonged are in the group known as the lycopsids. This is a group on a branch of the evolutionary tree between mosses and ferns. The lycopsids were the first plants to grow into proper trees and so must have had a root system capable of supporting the crown of a tree. The fossil roots of these ancient lycopsids differ from more modern roots in that there is a single central core of xylem and phloem surrounded by parenchyma, and an epidermis with root hairs.

By the time the ferns were producing roots, a different developmental pathway had evolved. In the majority of ferns and seed plants, the root tip consists of four different regions. A root cap at the tip protects the more delicate structures from damage as the root pushes through the soil. Behind this is the meristem where the cell division happens. In the centre of this meristem is the quiescent centre which is believed to control the timing of the differentiation (as opposed to division) of the dividing cells that surround it. The new cells produced by the meristem then become the elongation zone as the tip grows away from them. Having expanded in the elongation zone, the cells then find themselves in the differentiation zone, where they develop into epidermis, cortex, endodermis, or vascular tissue. Further back, a tube of lateral meristem may be produced in perennial plants. This cambium will give rise to secondary xylem (on the inside of the tube) and phloem (on the outside), thus resulting in bigger permanent roots. These roots then will branch and form the complex root systems seen when high winds uproot mature trees. While this type of root system is common, there is another type known as adventitious roots. These grow from a part of the plant *other* than the primary root that grows from the embryo in the seed. This is the situation in the monocots and accounts for why mature palm trees can be transplanted with very little root. This will be explained in more detail in Chapter 5.

So now we have a diverse range of land plants ranging in height from ground-hugging liverworts to lofty coastal redwoods, and from tiny floating duckweeds to colossal eucalyptus trees. However, no matter how big or small, they share one goal in life: to make more plants for after they are dead. This they do on many variations of the previously described alternation of generations. Time for Chapter 3.

Chapter 3
Making more plants

The production of more new individuals resembling their parents is one of the defining features of living organisms and plants are no exception to this rule. When plants first colonized land and air, they had a reproductive system in place that worked in water, but could it still function where water was not ubiquitous? Obviously, the answer to this is yes, but some things had to change.

There are two fundamentally different ways to make more plants that we see in most, if not all, of the major groups of land plants. The more simple way is to make a replica vegetatively. The advantages of this method are that it is simpler than sexual reproduction, only one plant is required to make more, it is quicker, and if the parent is well adapted to its environment, then so will the offspring be. We find vegetative (or asexual) reproduction as a common trait of invasive species, so it is clear that there are powerful, short-term advantages to this strategy. There are a number of ways in which plants can produce free-living replicas of themselves.

Asexual/vegetative reproduction

Liverworts have a structure known as a gemmae. This is an undifferentiated blob of cells that grows in the gemmae cups on the surface of the thallus. When the blob is ready to be released, it is dislodged by raindrops and is washed away. When it comes to rest, it grows into a plant like its parent. Mosses like *Aulacomnium androgynum* produce gemmae from the tips of their shoots. These gemmae are not amorphous blobs but have distinct ends with meristematic apical and basal cells with several cells in between. The dispersal of these gemmae is by any means available.

Anyone who has been fell walking in the English Lake District should not fail to be impressed by the ability of bracken (*Pteridium aquilinum*) to spread. As the plant grows across the hillside, the individual shoots are connected to the parent plant. However, these shoots each have their own root system and are quite capable of living on their own. A few other ferns produce tiny plants from their leaves. *Asplenium viviparum* and *Asplenium bulbiferum* are two such species.

Vegetative reproduction is relatively rare in gymnosperms, but a well-known example is found in the coastal redwood (*Sequoia sempervirens*). This species produces young plants in a ring around its base. When the parent plant dies, it leaves a concentric ring of young trees. Eventually, these will each produce another ring of replacements. It could be argued that this is regeneration rather than propagation of the plants, as there is no hint of dispersal. The Huon pine (*Lagarostrobus franklinii*) from Tasmania also produces these long-lived clonal populations which have been claimed to be more than 10,000 years old and so are the oldest organisms, or genotypes, on Earth.

As all gardeners will know, vegetative reproduction is very common in flowering plants. Couch grass (*Elytrigia repens*), ground elder (*Aegopodium podograria*), and bindweed (*Convolvulus arvensis*) are just three plants that are very successful weeds of cultivated places because of their ability to regrow into mature plants from just a short fragment of plant that is left in the soil following digging. As mentioned before, many of the world's most invasive plant species take over an area by rapid vegetative spread. Plants in this category include *Rhododendron ponticum* and brambles. In the former, the lowest branches will root as they touch the soil, and in the latter the shoot tip will develop roots if it is in contact with soil for a few weeks. A particularly insidious version of this form of reproduction is found in false garlic (*Nothoscordum inodorum*) which produces tiny bulbs, or bulbils, from around its main bulb and also near the flowers in the inflorescence. This, combined with prodigious seed production, makes this an impossible plant to eradicate from where it is not wanted. The water lettuce (*Pistia stratiotes*) produces perfect little plants on the ends of lateral shoots. These plants are broken off by disturbance by large animals. However, if left undisturbed, this species will form huge mats that can become floating islands strong enough to support a rhinoceros (albeit a small rhinoceros).

7. *Nothoscordum* produces many small bulbils from around the parent bulb, making it a very persistent weed

In some plants, there is a strange half-way-house between sexual and asexual reproduction known as apomixis (or sometimes parthenogenesis, a term more commonly used when animals are being discussed). In apomixis, seeds containing a viable embryo are produced without the egg being fertilized. This means that the seeds are genetically identical to the mother plant. The problem for botanical recorders is that each region of the world may have slightly different clones of the species. This is the case in dandelions, for which there are more than 250 named micro-species that are simply clones. These clones may be short-lived, but there are implications here for conservation. Do we try to conserve each variant, or should we try to preserve the process by which this number of micro-species is maintained?

However, being well suited to a particular habitat today is no guarantee of future success if the habitat changes. While evolution is incapable of foresight, it can be seen that the ability to produce variation, through mutations and reassortment of genes during sexual reproduction, has provided the modifications that Darwin recognized as being central to evolution. As Darwin suggested, 'it is a general law of nature that no organic being self-fertilizes itself for an eternity of generations', because this leads to something known as 'inbreeding depression' whereby progeny become weaker and weaker. This being the case, we come to the second, more complicated, means of reproduction, and that is the fusion of a sperm and an egg in sexual reproduction.

Sexual reproduction

In Chapter 1, the life history of plants was introduced, and it was shown that this involves a strategy unique to plants – namely, the alternation of a diploid and haploid stage known as the alternation of generations. If we accept that the green algae like *Chara* are the closest relatives of the land plants, and if we accept that liverworts are the living descendants of the lowest branch on the evolutionary tree for land plants, then the differences between these two groups of plants are a clue as to what changed to enable plants to colonize land and air. In Chapter 2, we have seen how many of the problems were solved (desiccation, support, finding raw materials and nutrients), but there is one outstanding problem, and that is how to bring gametes together when the previous means of swimming through the omnipresent water is not always an option.

To recap very briefly, the *Chara* plant produces lots of sperm with two tails, or flagella, that swim to another plant where lots of eggs are sitting in a structure known as an oogonium. The sperm fertilize the eggs (one sperm per egg) and a zygote results. The zygote then undergoes meiosis to produce four spores. Sperm, eggs, and spores contain one set of chromosomes (haploid), whereas the zygote has two sets (diploid). There is one very big difference between this life history and all of the land plants, and several smaller differences. The very big difference is that the zygote of land plants does not immediately divide meiotically. Instead, it divides mitotically (simple cell division) and grows into an embryo. This is new, and this is why the land plants are referred to as the embryophytes (by botanists mainly!).

Mosses and liverworts

So we start with the liverworts, and in particular with *Marchantia polymorpha* as this is probably the most studied and best understood liverwort. The plant of *Marchantia* is a dark-green thallus, shining like wet liver and bifurcating as it spreads across the surface of the moist growing medium. On its surface, there may be the gemmae cups, each about 5 to 8 millimetres in diameter. However, at certain times of the year, wet times, umbrella-like structures growing from the surface of the thallus can be seen. There will be some umbrellas that have an intact top and some that will look skeletal as if all the material has blown away and only the spokes remain.

From the upper surface of the intact umbrellas, sperm will swim from the antheridia. These antheridia (singular, antheridium) are one of the smaller differences between *Chara* and the land plants, though they don't *look* very different from the sperm-producing organs of *Chara*. The antheridia are one of the two types of gametangium found in land plants. (The suffix -angium, plural, -angia, is used to describe a reproductive structure, so gametangia produce gametes and sporangia produce spores.) The other type of gametangium is the one that produces the egg – the archegonium; the archegonium is the other small difference between *Chara* and the land plants, though they don't *look* very different from the egg-producing organs of *Chara*. A significant difference is that the archegonium produces just one egg, while each oogonium of *Chara* produces lots of eggs.

So the sperm swims using its two flagella towards an egg in an archegonium. These archegonia are on the underside of the umbrellas with just spokes. The plants will be either male or female and so just one type of umbrella. The mechanism that determines the sex of these plants is genetic and is due to heteromorphic chromosomes: a small Y (male) chromosome and a large X (female) chromosome. This might sound familiar, and indeed in humans males are XY and females XX, but you must remember that the liverwort plants producing gametes (the gametophytes) are haploid, so the males are just Y and the females just X. Meanwhile, back at the liverwort archegonium, the egg has been fertilized by the sperm and a diploid zygote formed. This zygote now divides and grows into an embryo. It is *not* released by the archegonium but derives its nutrients from the female gametophyte. The embryo grows and develops a sac hanging on a stalk; this is the sporophyte of the liverwort. It is not free-living but dependent upon the gametophyte for all its 'life'. Inside the sac, spores are being produced and wrapped in sporopollenin. This is the tough outer layer described in Chapter 1. The sac is therefore the sporangium. The spores are produced following meiotic divisions that reduce the number of chromosomes from diploid back to haploid. The spores all *look* the same, and so this plant is said to be homosporous. In reality, the spores are not all the same because half are female and half are male, but they are the same size and are produced by the same sporangium.

These spores fall from the ruptured sporangium and are blown away by the wind or washed away by water. The spores are tough little structures because of their coat of sporopollenin that they inherited from their aquatic ancestors. When the spores come to rest on a suitable medium and the temperature and humidity are right, they germinate. They grow into either a male thallus or a female thallus, and these gametophytes then grow until mature, when they will put up their umbrellas and the life history is complete.

The life history of the mosses is similar but different. We already know that mosses are one step ahead of the liverworts because they have stomata. We may be more familiar with mosses as they are common plants, in towns growing on walls and in gardens growing in lawns. The plants growing on walls are more useful to consider here because at certain times of year, particularly the late winter/early spring, the plants sport little structures that look like periscopes – a thin filamentous stem supporting a capsule a few millimetres long. This capsule is the sporangium and so out of its open end come tough haploid spores, and again they all look the same. We are ignorant about the determination the sex of the gametophytes of mosses, but in *Ceratodon purpureus* it has been shown to be genetically determined probably by sex chromosomes. However, in some mosses like *Physcomitrella patens*, the spores germinate and grow into a bisexual (or hermaphrodite) gametophyte.

Irrespective of whether the gametophyte is bisexual or single sex, antheridia and archegonia develop, often near the tips of the leafy shoots. When the weather is wet, the sperm swims from its antheridium to an archegonium, where it hopes to find the egg of its dreams. How it finds the archegonium is uncertain, but in some species there appears to be a chemical signal being emitted by the archegonium. Once the sperm has fertilized the egg, the zygote grows into a multicellular structure that grows into the sporophyte. In the case of the *Tortula muralis*, that sporophyte is the periscope that we see on the tops of walls in late winter. Within the capsule, meiosis takes place, haploid spores are produced, and the life cycle is completed.

Ferns

The next major group of extant plant species on the tree of life are the ferns. If you look at the underside of the older leaves of ferns, you will often see erumpent pustules releasing a brown dust. If you cut the leaf off the plant, secure it gently on to a piece of white paper, and put the paper on top of a boiler or Aga overnight, when you remove the fern leaf the following morning, you will have a perfect representation of the distribution of the sporangia on the fern leaf, because that brown dust is thousands of identical spores all coated in sporopollenin. This means that the erumpent pustules are the sporangia and the leaf is part of the sporophyte. This is obviously rather bigger that the periscope of our little moss growing on the top of the wall, and it is different in another way: this mature fern sporophyte is free-living, whereas the periscope moss sporophytes were completely dependant on the gametophyte underneath them.

So the spores are released and are blown away, hoping to land on a piece of wet soil or a wet branch, or anywhere in fact with reliably high humidity and 'soil' with some nutrients. The spores germinate and grow into the gametophytes. The gametophytes are free-living and green, but they are not large. They are rarely more than 10 millimetres in any direction. They are often flat and thallose, not dissimilar in many cases to a small liverwort, and they can be mistaken for liverworts since they often grow together in the same conditions. The gametophyte may be male, female, or bisexual depending on the species or on the conditions. In *Ceratopteris richardii*, the first spores that germinate are bisexual, producing both archegonia and antheridia. However, these pioneering gametophytes produce a hormone that affects the development of any newly germinated spores nearby, and if the hormone reaches these young gametophytes in first few days (two to four days), the gametophyte is totally male and produces just antheridia and thus just sperm. There is an advantage to this strategy in that it promotes outbreeding. Another way to promote outbreeding is for the gametophyte to produce the gametangia at different times, thus being male for a while and then female, or vice versa.

A problem inherent in the gametophyte producing both antheridia and archegonia is that it could produce just self-fertilized eggs and thus zygotes that are a diploid clone of the gametophyte. Among Charles Darwin's portfolio of biological one-liners is 'Nature abhors repeated inbreeding', and, as we shall see later in this chapter, Nature goes to a lot of trouble to avoid inbreeding – most of the time. 'Most of the time' because one of the challenges facing all plants, be they aquatic or terrestrial, is that they are immobile. If they are the only member of their species in the vicinity, they are not going to be able to produce any young at all, if they have to be fertilized by sperm from a different plant. So many plants have a policy of accepting their own sperm if there is no alternative, so that at least they get another generation and another chance to outbreed. Darwin called this strategy 'reproductive assurance', and it is widespread.

Leaving aside concerns about where the sperm have swum from (and they now have many flagella to help them on their way), a zygote is formed in the archegonium, and this grows into an embryo. However, rather than growing into a sporangium on a stalk of some type, the embryo produces a little leaf above the gametophyte and a little root below it. This young plant is initially supported by the gametophyte, but by the time it has a few leaves and a few roots, the parent gametophyte has been drained of all its resources and is an empty vessel – a possible metaphor for the human condition.

So far, all of the sporophytes in this story have produced just one type of spore; they have been homosporous. It is believed that fossil plants like *Cooksonia* were also homosporous. However, many plant groups and the vast majority of plant species currently extant are heterosporous. Heterospory has evolved several times, and the first time may have been before the ferns came to prominence. Between the mosses and the ferns on the tree of life are the lycopsids. This rather enigmatic and diverse group of plants was among the first to grow into trees and also the first to produce male and female spores. Under no circumstances must these spores be confused for gametes. They are spores, so like gametes they are haploid, but all a spore can do is grow into a gametophyte. Spores cannot fuse with each other to give a diploid zygote.

Seed plants

The last major group of land plants are all heterosporous, and these are the seed plants: the gymnosperms and the angiosperms. The gymnosperms include the conifers, the cycads, Ginkgo, and a small bunch of botanical weirdos including *Welwitschia mirabilis*, the iconic oddity from the Namib Desert, and *Ephedra distachya*, the original source of medically exploited ephedrine (described in more detail in Chapter 6). The angiosperms, on the other hand, are the flowering plants. It is believed that the seed plants have a common ancestry, and it is assumed that their most recent common ancestor was heterosporous, but what it looked like is the subject of several careers in scientific research and inspired guesswork. The emergence of seed plants may be a mystery, but the origin of the angiosperms is an 'abominable mystery' according to Darwin, and it is not yet solved.

8. *Lycopodium* in Japan. Relatives of this plant were among the first plants to grow into trees

Evolution is a complicated business. It has no foresight, so it simply retains what works today and does not have an attic full of things that might be useful in the future. It is also quite uncharitable. Darwin believed that no organism would acquire a trait for the benefit of any organism but itself. It may be that what is advantageous to one organism might be exploited by another at some point in the future, but that will not be considered by evolution. The seed is a good example of this. It is difficult to think of the world being as it is now had not seeds evolved on seed plants. Many animals, including 6.5 billion humans, depend on seeds for their staple food. Yet seeds can only have evolved if they gave the plants that possessed them an advantage. Those advantages might have been survival and dispersal, and perhaps the ability to survive dispersal. So seeds permitted transport in both space and time. But what is a seed, and how is it produced? Seeds have been described already in Chapter 2 as 'an embryo wrapped in a tough coat with a packed lunch in its pocket'. If the seed has an embryo in it already, then it is formed *after* reproduction has happened.

If you have ever parked your car under a cedar tree (*Cedrus* spp.) in the spring, you may well have returned to find it covered in yellow dust. You might even have found some soft, sausage-shaped cones on the ground around the car. If you look into the tree, you will see more of these small cones and some very big woody cones a few inches in diameter. Inside these woody cones is where this part of the story begins because here we find a sporangium. (It is in fact known as a mega-sporangium because it produces larger spores than the other type of sporangium found on a seed plant.) These spores are not released as has happened in every species we have looked at so far. The megaspore is retained in the cone, where it grows into a female gametophyte that is similar in size to the fern gametophytes. An archegonium grows on the gametophyte, and so again this is not so different from the ferns, except that this gametophyte is not free-living but is totally dependent on the sporophyte for its nutrition.

Now if this has made any sense, you should be asking yourself where is the male gametophyte? Well, the male gametophyte is made somewhere else and then has to find the female. The male gametophytes grow from microspores produced by the microsporangium and these sporangia are found in the other cones, the cones you found on your car bonnet. The microspores are produced with the traditional coat of sporopollenin. However, before they are released, the spore inside undergoes a bit of cell division, and in addition two large air sacs grow on the outside of the sporopollenin. This whole structure is an immature male gametophyte. It is immature because it has no means of dispensing its sperm yet. This immature male is hoping to find a mature female gametophyte. The problem is that rather than the two gametophytes sitting next to each other on the soil, as was the case with the ferns, mosses, and liverworts, the females are sitting in a woody cone elsewhere on the tree, or preferably on another tree.

The ingenious males find their females by gliding, buoyed up by their air sacs. The hope is that one of them will land on the woody cone, preferably on another tree. There is a million to one chance of this happening, but this is enough. These gliding immature male gametophytes are much better known as pollen grains. When it lands on a woody cone, the pollen grain takes in water from the imaginatively named pollen drop. It then grows a tube and at the tip of that tube is a sperm cell. This sperm is not produced from an antheridium; these are now a thing of the evolutionary past. This sperm has no flagella, so the tube has to grow to the neck of the archegonium on the female gametophyte that has been waiting patiently for its male caller. The sperm fertilizes the egg, and a zygote is produced that grows into an embryo; so far, so good, except that the embryo is nowhere near a suitable place to grow, so the development is halted, a thick coat is provided by a part of the woody cone, and the female gametophyte takes on the role of 'food for the journey'. The seed is released from the woody cone and floats to the ground, hopefully some distance from its mother. Once there, the embryo resumes its growth and grows into a cedar tree.

If this cedar tree had instead been an oak tree, or a magnolia, or a daffodil, the process of bringing the sperm and eggs together would have been different again. These flowering plants do produce pollen, but rather than the microsporangia being situated in a cone, they are in the anthers on the end of the stamens in the flowers. The pollen is similar to that of the cedar tree (and other gymnosperms) in that it is an immature male gametophyte on the look-out for a mature female gametophyte. The problem for the flowering plant's pollen is that the female gametophyte with its egg has been hidden inside the carpel. This carpel is the major innovation in the evolutionary transition from the gymnosperms to the flowering plants. The carpel consists of three parts: the stigma (the landing-pad for the pollen); the style (a stalk, of variable length depending on the species, that connects the style and ovary); and the ovary that contains the female gametophyte(s), each with its own egg. The female gametophytes have developed from the megaspores. These megaspores have been produced by the megasporangia in the ovary. It has to be noted that these female gametophytes are a sad apology for a gametophyte compared to everything that has been produced by other land plants. These angiosperm female gametophytes consists of just a few haploid cells (one of which is the egg cell that has been produced without the aid of an archegonium) and one cell that has two haploid nuclei. It is easy to make a case for this binucleate cell being awarded the prize for being the most important cell in the world, because it develops into not only food for the embryo, but it is the food for the world in grains of rice, wheat, and maize.

So the pollen lands on the stigma. How it does that with precision is one of the best stories in plant biology and will be dealt with at the end of this chapter. For now, we can assume that the immature male gametophyte has landed on the stigma, and its problems are just beginning because this is 'meet the mother-in-law'. The

54

stigma has two questions for this young male. Firstly, are you the right species? Some mothers are less fussy than others, or just visually challenged, and this is why sometimes pollen from another closely related species gets through and hybrids may be produced. This can be a very important way of producing novelty for evolution to exploit or reject. The second question is, are you me? As with all organisms, inbreeding is second-best for plants. Again, some plants are much more diligent in this respect than others. Plants in the cabbage family, for example, are very keen to avoid being fertilized by their own sperm. Once the pollen has answered the questions adequately, water is passed to it and the tube grows down the style. More than one pollen grain may have germinated, and there is evidence that there is now a race to the eggs and that selection for particular traits may happen at this stage.

When the pollen tube emerges from the end of the style into the ovary, it grows towards an egg and releases not one but two sperm. One of these does the decent thing and finds an egg and makes an honest zygote out of it. This zygote grows into an embryo that then has its development arrested, as was the case in the gymnosperms. The second sperm, however, is not just a spare. This has to find the cell of the female gametophyte that has two haploid nuclei. The sperm fuses with this oddity, with the result that the cell is triploid, that is, it has three copies of each chromosome. The reason why this might be considered 'the most important cell in the world' is because this goes on to grow into the endosperm, which is not only the food supply for the embryo but also the food supply for *Homo sapiens*. When we eat anything made from rice, wheat, or sweetcorn, we are eating triploid endosperm.

So how does the pollen of flowering plants get moved with precision? Well, in some cases, it does not and wind is employed with the usual long odds. Rarely, water is used, but more often an animal is hired if a suitable reward can be provided. The rewards are varied. Provender (nectar, starch, and pollen itself), a bed for the night, a bed for the night with someone else, somewhere to lay your eggs and to raise your young are the rewards on offer. The size of the rewards vary depending on the size of the visitor, so a bird or bat requires a bigger reward than a midge. Attracting the pollinator is important, but animals are straightforward and so attracting them is simple. Smell, shape, and colour are all used. Some plants have become embroiled in very close relationships with just one pollinator, which is fine so long as nothing happens to the pollinator. Some species do not care who comes calling just so long as someone does. This is going to increase the risk that the pollen lands on an inappropriate stigma, but at least the plant has been visited.

It is often possible to match pollinator to flower and flower to pollinator, and the study of pollination syndromes was the part of Charles Darwin's experimental work that gave him the greatest pleasure. He wrote a great deal about the various contrivances that plants employ. However, it has recently been shown that this, like so much of plant biology, is not as precise a relationship as one might assume. Some seemingly generalist flowers with no specific adaptations are still only visited by one pollinator, and conversely some seemingly well-adapted flowers are visited by many different organisms. Likewise, although it is possible to describe your perfect bee-pollinated flower or bird-pollinated flower, these are rarely found in nature when you look for them.

So pollination, or the distribution of immature male gametophytes, is a critically important stage in the life of seed plants. It helps to stir up the gene pool as well as to distribute genes. In the seedless plants, the flagellated sperm has to take on this responsibility. As this has worked for more than 450 million years, it can hardly be seen as a problem. However, there is another problem for which neither a flagellated sperm nor a pollinator is suited, and that is the dispersal of the new embryo away from its parents. The seed plants have seeds but the seedless plants have another way, and this will be looked at in the next chapter.

Chapter 4
Moving around

If you live in the water, there are two ways to get around: drift or swim. Some of the glaucophytes are swimmers, some of the red algae are drifters (such as *Polysiphonia*), and some of the green algae are swimmers (such as *Chlamydomonas*, *Volvox*, and *Ulva*) and some are drifters (such as *Spirogyra* and *Chara*). The swimming referred to here is that of spores (*Ulva*) or the whole organism (*Chlamydomonas* and *Volvox*). The swimming of sperm does not count, as this is not vegetative dispersal, it is genetic dispersal of the male germ line but nothing more. A sperm cannot develop into a free living organism. The swimming spore of *Ulva*, on the other hand, is able to grow into a free living plant. Is dispersal of plants that live in water important when compared with land plants? Is dispersal important for land plants? The answers to these questions seem to be coloured in some societies by the human condition. It is assumed that children will leave the family home in some regions, while elsewhere they are expected to join the family team and support the group. Plants are asentimental, and dispersal mechanisms will be selected for only if they increase the chance of the plant's genes getting into the next generation.

So back to the questions. Is dispersal of plants important? The aquatic environment is more homogenous than the land. The abrupt barriers that exist on land are rarer in water; gradients exist but they tend to be gradual. Swimming can be useful in this type of environment because it enables young plants to get away from other members of their species with whom they will be competing more strongly than any other organism. However, and this applies to land plants just as much, if your parents are growing well in one place, why risk leaving the area? Perhaps this is why as many as 80% of land plant species simply drop their seeds at their roots and make no attempt to promote their dispersal. Compared to the aquatic environment, land is spatially heterogenous, and so we might expect to find more land plants investing in dispersal mechanisms. However, in addition to spatial heterogeneity, land can be very temporally heterogenous, so we need to consider the dispersal of plants in time as well as space because plants can enter suspended animation in a way that animals can only dream about.

Let us first consider the liverworts and mosses. These plants produce haploid spores that grow into free living gametophytes. We see these plants not only at ground level but also on trees, on high roofs and walls and occasionally in the window ducts of old Land Rovers. The spores are light enough to be blown to any height and, thanks to their coat of sporopollenin, to survive the journey. The same is true to a certain extent for ferns. There are many ferns that do not live in soil. There are epiphytic ferns that grow on the outside of branches, some are free-floating and some grow between the stones in walls. For the epiphytic species, the spores must be able to reach the branches of the trees upon which they are to grow. This is equally true for epiphytic flowering plants such as orchids, many bromeliads (the pineapple family), and some members of the arum family, but more of these later. So from the most casual observations, liverworts, mosses, and ferns seem to have no trouble surviving and spreading. Furthermore, the spores are so small that they are not going to be a very attractive food source. This cannot be said for seeds, so what is so special about seeds? Why would evolution favour the production of such an energy-expensive structure as a seed that would become the favourite food of many organisms?

There are very few, if any, rules in seed biology. There is still a great deal of research to be carried out before we can say that we understand the ecology of seeds. One problem is that there are at least 350,000 species of seed plant and they are all different. Species within a genus can differ in their seed biology, and even plants in the same species can vary in different localities. For example, plants of *Pinus bruttia* grow from Crete to northern Greece. At the south of its range, it never experiences frost, whereas up north nights frequently fall below zero in the winter. The seeds of the southerners show no dormancy and will germinate when the autumn rains begin, like many plants in Mediterranean-type regions of the world. On the other hand, the seeds from northern plants do not want to germinate just before a harsh winter, and so their seeds require chilling before they will come into growth in the spring. These are still the same species because acclimatization is not the same as speciation.

One thing we can state without fear of contradiction is that seeds are much smaller than their parents and that they are able to withstand harsh conditions that would kill their parents. It is this ability, perhaps more than any other, that influences the various strategies seen in plants that produce seeds. For those plant species that flower once in their life and so produce just one batch of seeds (monocarpic species), it is critically important that at least one of their seeds survives to flowering size to take their genes to the next generation. It will probably be stated many times in this chapter, but it bears repetition, that the vast majority of seeds do not result in a plant that produces more seeds. If you are an oak tree and have many seasons of seed production, this is not an issue, but for an annual you only have one chance.

Each of the 350,000 species of seed plants is different from the others. Each individual within a species has choices to make during its lifetime, leading to even more diversity. Biologists sometimes try to reduce biology to the allocation of resources in order to rationalize the behaviours and actions of living organisms. So when plants produce seeds, they have to balance the need to grow with the need to put energy into the production of those seeds. It is clear to see that it takes a lot out of a plant producing seeds because in those species where the male and female flowers are on separate plants, the females tend to be smaller and shorter lived than the males.

Do you produce many small seeds or fewer bigger seeds? If you produce lots of seeds, does this increase your chance of survival? The fact that monocarpic species tend to produce very large numbers of seeds implies that the probability of survival increases the more seeds that you produce. Do you use fats or carbohydrates as the energy supply in the seeds? Fats are more expensive to make but are richer in energy gramme for gramme. Plants put between 2.3% and 64.5% of their resources into seed production. There are very few rules in seed biology.

Another issue is how long to take to reach maturity. Unlike animals, sexually active plants of the same species can and do vary greatly in size. If highly disturbed, plants tend to rattle though their life history fast before the next cataclysmic change. In more stable places, seed production may be constant. For example, fig species, which all have a unique relationship with a different species of fig-wasp, produce flowers all year round to satiate the wasp. Beech trees, on the other hand, demonstrate a different strategy, that of masting, whereby a very large number of seeds is produced approximately every seven to ten years. It is believed that the advantage of this is to produce so many seeds that there are far more that the natural predators can consume. Oak trees, on the other hand, produce many acorns each year, many of which get buried by squirrels and then forgotten about. This appears to be good news for the oaks.

9. *Lodoicea maldavica* has the largest seeds in the world. It grows only on the Island of Praslin

Seeds vary greatly in size. The seed of orchids is tiny, almost dust-like, whereas the seeds of the double coconut, *Lodoicea maldavica*, is the size of two Aussie Rules footballs; that is, a variation of seven orders of magnitude in weight, and yet both the orchid seed and the coconut seed fulfil the same function, and both have the potential to produce a mature flowering plant. Even on the same plant, seeds can vary in size by three orders of magnitude, and the seeds within the same fruit can be different. One rule of thumb seems to be that plants that live in shade tend to have seeds from the bigger end of the scale, perhaps as this enables them to become established in an environment with lower energy levels. Seed size appears to be linked to neither the moisture content of the soil nor the nutrient status of the soil. It may just be that by producing seeds of various sizes, the plant is hedging its bets and increasing its chances. One size of seed does not fit all situations.

In the previous chapter, we saw how many plants have come to rely on a stranger to distribute their sperm, and those who do not trust an animal have thrown caution literally to the wind. We know that neither biotic nor abiotic pollination methods are totally satisfactory because hand pollination of a plant is nearly always more successful than leaving it to nature. Some families of flowering plant are particularly hopeless in this respect. Only up to 7.2% of flowers in the Protea family are successfully pollinated, though far more are visited by the pollinator. It may be that their chosen pollinator is now extinct. (Plant species have a general problem when forming joint ventures with animal species, and this problem is that plant species last 30 times longer in evolutionary terms than animal species. So if you are going to use an animal to distribute your pollen or your seed, you will have to be prepared to change suppliers repeatedly.)

We see in plant reproduction a good example of the Allee effect, namely that it is better for the members of a species to hang around with each other – the scientific safety in numbers principle – because the fewer plants and flowers there are, the fewer successful pollinations there are, and so the fewer seeds are produced. Before you start to think there is a rule appearing, you have to remember that although the bigger groups of inflorescences attract more pollinators and so produce more seeds, they also attract more flower grazers, more seed predators, and more fruit grazers. So there is no optimum size for seeds; the best size is time and space specific, and so the best strategy is variation.

Charles Darwin was not only interested in pollination, he was also fascinated by the geographical distribution of species. He carried out some elegant experiments at Down House to discover if seeds of land plants could survive the salinity of sea water for long enough to enable them to float from one continent to another. He suggested the feet of birds as another means of long-distance dispersal, particularly to oceanic islands, and this is one of the first things that children learn about in school nature lessons. They are shown how the fruit of goosegrass (*Gallium* spp.) and teasels (*Dipsacus fullonum*) adhere to the fur of animals, and because it is the first thing that we are shown it is assumed that this is how most plants disperse their seeds. The truth is that fewer than 5% of species travel on the outside of animals. These species are low-growing (that is, animal height) and from many different habitats. Wind dispersal is also assumed to be common after we have dropped a sycamore 'helicopter', and yet very few structures improve the lateral movement of seeds, they just make them fall more slowly, allowing the horizontal wind to have more effect. Wind can lead to extraordinary dispersal if it is very strong, but it has been shown that generally a forgotten cache of seeds will be further from the parent plant than a wind-dispersed seed. So animal fur and wind are responsible for aiding seed dispersal but in a small minority of species. Some parent plants physically eject their seeds using a ballistic mechanism. The most spectacular of these may the squirting cucumber, *Ecballium elaterium*, that releases its seeds as a stream of exhaust as the fruit breaks off from its stalk. This is so efficient that every seed is ejected during the short flight of the mini-cucumber.

The main alternative to these two is the seed being eaten by an animal and being excreted by that animal some distance from the mother plant. This is a risky strategy as the animals (normally a mammal or bird) have eaten the seeds for their nutritional value and they will grind up and kill the seeds in order to extract that food. Interestingly, those seeds that are good at surviving a trip through the gut of sheep are also those species that persist for many years in the soil seed bank. One way to reduce this risk of being digested is to coat the seed in a laxative; this is more common in bird-distributed species. This may be a good strategy for the plant, but it does make the study of bird dispersal difficult because the dispersal is random and difficult to map! In the same way that pollinators and flowers can be paired up, so can seeds and animals. Birds are attracted to scentless, brightly coloured seeds – reds, blues, and blacks are often combined – whereas mammals go for something smelly and tasty but dull-coloured. The actual evidence to support the dispersal of viable seeds by animals is thin. In two studies, 40,000 deer pellets and 1,000 kilograms of rhino dung were inspected and no viable seeds were found. In another study, of 40,025 seeds eaten by house finches, only 7 survived the experience.

Evidence for the impracticality and foolishness of using the gut of an animal as a vehicle can be found in the fact that many seeds and fruits are toxic. As with many aspects of plant biology, there are many reasons why a seed or fruit might benefit from being toxic and they are by no means mutually exclusive. A tasty but laxative fruit will ensure swift and safe passage of the seed through the gut. An emetic may ensure swift and safe passage of the seed out again before it reaches the gut. If toxic, it might kill the animal, giving the seedling the nutrients of the decaying host. The toxins might be there to inhibit the germination of the seed – physiological dormancy, of which more later. The toxin may be

10. Seeds of *Euphorbia stygiana*, showing the elaiosome

toxic to seed predators but not to fruit dispersers. A good example of this is the English yew, in which the red aril is safe to eat but the seed is toxic. Finally, the toxin may be a defence again pathogens.

One group of animals that are important in seed dispersal, particularly in the Mediterranean-type regions of the world, are ants. Ant-dispersed seeds possess an elaiosome. This body of fat, usually much smaller than the seed, is an intoxicant of the ant in so far as it stimulates the ant to pick up the seed and take it back to the nest, where it bites off the elaiosome before taking the food inside the nest, leaving the intact seed outside. The soil around the nest is often higher in nutrients than nearby, and often the seeds are taken a couple of inches below ground, where they are protected from not only other predators but from the harmful effects of fire that is common in these areas. Many species of plant are dependent on the ants for this protection, and so when non-native ant species drive out the native species, the plants are left exposed and vulnerable to local extinction.

The problem of non-native species will be dealt with in more detail in Chapter 7, but it is relevant now because the main reason why plants grow here and not there is that they cannot get there. Humans, however, have broken down every dispersal barrier, accidentally and on purpose. Farmers have moved seeds around in contaminated seed sacks, in manure, and on stock – 400 sheep have been shown to move 8 million seeds per annum. Gardeners have moved thousands of species well beyond their natural ranges. Finally, foresters have introduced more productive trees into many countries. The damage that these non-native species can do must not be under-estimated.

Seed dispersal is 'good' for several reasons. Firstly, the parent plant will have become a mecca for pest and diseases, and very few diseases of plants survive in or on a seed. This predator and disease release is one reason why plants in a new country can do so well. Secondly, spreading your seeds reduces the risk from stochastic, or sudden and unpredictable, disasters. Thirdly, it reduces parent–child competition (more anthropomorphizing of botany). Finally, it increases the chance of finding a 'safe site'. Yet in reality, more than 80% of plants make no effort whatsoever to distribute their seeds, so having no adaptations for seed dispersal is no disadvantage until you have to migrate fast. If the climate is going to change fast, then fast migration may be selected for very positively in the next few decades and centuries.

Seeds are dispersal units, and by their ability to survive in soil seed banks they can be said to be dispersed in time. It is difficult to find any generalizations about which species of plant and which types of seeds you will find in a soil seed bank, partly due to the difficulty of finding the seeds. Generally, though, small seeds persist longer in the soil as they are less attractive to predators and because they more easily fall into the cracks in the soil than large seeds. Soil can hold many seeds if they are small enough, the record being 488,708 per square metre. The record for a seed surviving in soil is held by a seed of the sacred lotus, *Nelumbo*

11. The seeds of *Nelumbo nucifer* remain viable for over 1,000 years

nucifer, which germinated after 1,288 years (give or take 250 years). In the UK, an experiment was initiated by W. J. Beal in 1879. Viable seeds are still being extracted from the soil more than 120 years after they were buried.

Seeds in the soil seed bank may not be dormant, they may just be in the wrong conditions for germination. In this state, the seed is said to be quiescent, and for buried seeds light is what is missing. In order to persist, the seed may be dry, but some seeds survive better imbibed as they are then better able to repair routine damage inflicted on membranes and DNA. However, the longer that a seed remains buried, the greater the chance of death from pathogens, predators, or premature germination followed by grazing. The ability to persist appears to be a trait belonging to a species rather than individuals within a species. Persistence is also related to the potential risk of totally failing to get at least one plant into the next generation. This means that plants in stable communities like woodland show less persistence in the soil's seed bank than seeds in Mediterranean-type habitats.

The difficult concept of dormancy has already been mentioned and needs to be properly understood. A seed does not germinate in a packet of seeds because the minimum requirements for germination are absent. These seeds are quiescent. This prevents the seed from germinating if the conditions are wrong at a moment in time. Dormancy is a state of the seed and not the environment. If a seed does not germinate within one month of being sown, then it is either dead or dormant. If it is dormant, then that dormancy has to be broken for the seed to germinate. The role of dormancy is to time germination when the *seedling* has the maximum chance of survival. Dormancy is less important in habitats that have uniform and predictable conditions, as these conditions alone will prevent seeds from germinating. Most species in tropical moist woodland do not produce dormant seeds.

There are three basic types of dormancy. Morphological dormancy is when the embryo in the seed is immature when shed, such as in orchid seeds. Physical dormancy is when a hard seed coat prevents the uptake of water and keeps the embryo dry, such as seeds in the pea family. Physiological dormancy is when some chemical change is required, such as in those seeds of the rose family that require a period of chilling. The first two are non-reversible while the latter is, thereby permitting some flexibility. Morphological and physical dormancy never combine; morphological and physiological often combine to give morphophysiological dormancy (MPD); and physical and physiological rarely combine.

Dormancy can trap a plant in a particular habitat, making it particularly vulnerable to changes in climate should they occur. However, the conditions experienced by the parent plant can influence dormancy, and so plants of one species can demonstrate different dormancy states depending where they grow. For example, trees of *Pinus brutia* that grow on Crete show no dormancy and germinate in the autumn when the rains come, while seeds from trees in northern Greece require a period of chilling to prevent germination until the winter is over and the spring conditions have arrived.

A commonly held view is that physical dormancy imposed by a hard seed is broken down by a period of abrasion or attack by a microbe. There is no evidence to support this view. The breaking of this type of dormancy is much more controlled, with a variety of different structures providing a temperature-operated one-way valve that permits the ingress of water. This strategy has evolved in several different groups of plants independently. Another strategy that has evolved several times is parasitism, and seeds of parasitic species perform the very neat trick of perceiving when their host plant is nearby.

The imbibing of water is the first stage of germination. This is followed by a rapid increase in respiration and the mobilization of food reserves. The embryo starts to grow, and when the young root emerges, the seed has germinated and there is no going back. Germination can be stimulated by a number of factors. Fluctuation of temperatures between night and day can be a means by which a seed perceives a gap in the canopy above the seed. Likewise, the ability to perceive the quality and quantity of light falling on a seed will enable the timing of germination to be coordinated with a specific time of year, a particular depth of burial, or the degree of shade. Light sensitivity is imposed by the seed coat, and it is worth remembering that the seed coat is provided by the maternal plant. This is one way in which the conditions experienced by the parent plant can affect both the dormancy and the germination requirements of the seed.

Availability of water will be a critical requirement for germination. Not all seeds are dry and tolerant of desiccation. At least 7.4% of all species that produce seeds are recalcitrant, meaning that they will not tolerate being dried out. Soil nutrient levels are important for the germination of some seeds. High levels of nitrogen are needed by weed species. A flush of nitrogen often follows the creation of a gap in the canopy, so light, temperature, and nutrient status of the soil can be interconnected. A widespread mechanism for breaking dormancy and promoting germination is smoke, and this is even found in species that do not currently live in fire-dominated habitats.

It is often stated that the leaf litter from dominant plants can inhibit the germination of seeds. Although the idea of suppressing one's offspring is attractive, this form of allelopathy has never been demonstrated in a natural system. Leaf litter can provide a mechanical barrier to germination and may keep light-requiring seeds in the dark. The effect of leaf litter on germination and seedling establishment is a classic biological puzzle because, while litter can provide a good seed bed for germination, it can also be a habitat for small, grazing animals such as slugs and snails.

So germination is affected by temperature (actual and relative), water availability, nutrient levels (especially nitrogen), fire, and the presence of a soil seed bank. It is suspected that all of these will change with a change in the climate. The implications of climate change on seed behaviour could be profound.

Even when the seed has germinated, it is still in peril. Studies have shown that more than 95% of seedlings may be eaten or otherwise killed. This is on top of all the seeds that were eaten or fell in inappropriate places. These losses still may not be show-stoppers. The absence of a suitable gap may be the biggest single reason for the failure of a seed to grow into a mature plant. John Harper introduced the concept of *safe sites* where all the conditions are present for germination and successful establishment of the seedling. These sites may be very small, and perhaps just 1% of seeds ever find a safe site. The soil must be correct, the mycorrhizal fungus must be present and will form its association within seven to ten years, and light levels must be appropriate.

Seed biology may appear to be imprecise. We know far more about seeds today than we did twenty years ago, but there is still an absence of any rules. Is this because the production of seeds is such an evolutionarily risky strategy? The embryo is a vulnerable stage in any life cycle, but to release your embryo with the added attraction to predators of a food supply seems like odd behaviour for any parent, and yet it works against all the odds. It has been wisely suggested by Fenner and Thompson that there is 'a regeneration lottery that results in the maintenance of species diversity by default'. This lottery has resulted in more than 400,000 species of plants. How can this diversity be organized to aid our understanding of biology? Chapter 5 has an answer.

Chapter 5
Making sense of plant diversity

The naming of plants is the oldest profession. If you go back to the Book of Genesis, chapter 2, verse 19, God brought the products of his Creation to Adam and suggested that He had done His bit and that it was Adam's turn to name the results. Irrespective of your take on the verisimilitude of this historical account, it does show that the author of the Bible recognized the need for plants to have names if he was going to tell his story. Adam did the job very quickly because it was just three verses later that God made him a plaything for an idle moment. Adam was able to do the job quickly because he was starting with a clean slate and there was only him.

Today, the situation is far more complicated because plants have names given them by many indigenous populations without reference to anyone else. So ox-eye daisies in Buckinghamshire are called moon daisies in Oxfordshire on the other side of the Chiltern Hills. In Germany, there are 108 names for the white water lily. The word 'lily' is itself widely applied to plants such as water lilies, arum lilies, lily of the valley, and Madonna lilies. Only the latter is a true lily. There is an advantage to these vernacular names. They are in the local language and therefore more easily remembered. This is very important if the name is deadly nightshade. This name is, however, meaningless in France, where the same plant will have a different name. The problem that we have is that in a world where

12. Ox-eye daisies have many different English names depending on where they live

people travel around and talk to each other, there must be a universally accepted set of names. Linnaeus, the great global cataloguer of plant names, stated that 'without permanent names there can be no permanence of knowledge'.

Thanks to Linnaeus (and others), we do have permanent names and they are in the form of a Latin (or Greek) binomial. The great thing about Latin and ancient Greek is that they are apolitical languages so no one can complain on nationalistic grounds. They are also no longer evolving as languages. If you know some Latin or Greek, then you can start to appreciate the descriptive nature of plant names. *Calliandra haematocephala* is a plant in the pea family whose flowers are crowded together in a hemisphere. Each flower is superficially made up of nothing but stamens and so the inflorescence looks like a powder-puff, hence its common name. The Latin name means 'beautiful male parts', which is as good a name as powder-puff, but it does make *Calliandra* an inappropriate name for a girl, and yet .... One does feel some sympathy with the Chinese because so many of their plants are named after foreigners. *Mahonia fortunei*, for example, is named after a Pennsylvanian nursery man and a Scottish plant hunter.

13. *Calliandra haematocephala* with flowers that appear to be made up of many male stamens, hence its Latin name meaning 'beautiful male parts'

There are now rules that dictate how a plant should be named legitimately. Firstly, you check that is does not already have a name. When you are sure that it is a new plant species, you give it a name in Latin and describe it in Latin, though in 2011 it was decided by the International Botanical Congress that this latter condition should no longer be mandatory. Secondly, you make a dried pressed specimen and place this in a herbarium, where dried and pressed plant specimens are stored in perpetuity. Finally, when you publish the name and description of the plant in a peer-reviewed journal, PhD thesis, or similar, you state which herbarium holds the specimen so that botanists can return to see the actual plant that you named for at least 350 years. While there are rules, there is insufficient vetting of the names that are published, and so for every species of flowering plant there is on average two or three legitimate names. These are referred to as synonyms, with the most widely used name usually being the accepted name.

If you talk to botanists from around the world, you discover that many naming systems exist and that they are never random allocations of names, like children's Christian names. In China, for example, they have named all their magnolias in one group known as *xin ji hua*. In New Zealand, *Cordyline fruticosa* is known in Maori as Ti Plant and the varieties are known as Ti this, Ti that, and Ti the other. Binomial nomenclature is a common feature of many vernacular naming systems. As people named the plants that they saw around them, they intuitively placed similar species in groups. We see this in the book credited as being the beginning of modern botany – *Enquiry into Plants* by Theophrastus, published about 2,300 years ago.

Theophrastus records his observations of plants, and in the process puts them into groups on the basis of what they look like. For example, he describes a bunch of plants that have flowers all emanating from a single point, with much divided leaves that clasp the stem at the base of their stalk. These are all umbellifers or members of the celery family (the Apiaceae). This book was still

14. The flower head of a hogweed plant. This belongs to a group of plants known as the umbellifers, and this group has been recognized since the 3rd century BC

being published as a textbook in 1644, and translations and facsimiles are still available. The study of botany continued through the Greek and Roman empires, and in the middle 1st century AD. Dioscorides published *De Materia Medica*. This work remained in use until the 17th century. It described the plants used medically by Dioscorides, but the plants were grouped by their morphological characters as well as medical properties. (We shall see later that these two often go together.)

With the Renaissance in Europe and the exponential growth in travel came a renewed interest in plants for their own sake rather than just for their utilitarian value. Many herbals were produced in the 15th century and onwards, and these books are very good examples of a utilitarian classification. We still see these informal groups. In our gardens, we group plants on the basis of the conditions that they need (rock garden, water garden, woodland gardens, and so on) or on their colour (the red border at Hidcote Manor, or the White Garden at Sissinghurst Castle). We have vegetable gardens, herb gardens, and fruit orchards. In other societies, we see groups of edible plants, fibre plants, poisonous plants, and so on. These classifications are obviously helpful and will never die out, but they are not all-inclusive and they are not unique, so one plant may appear several times.

The early botanists, such as Swiss polymath Conrad Gesner (1516–65), set out to describe all of the natural world they saw around them. Gesner was a meticulous worker, and the annotated drawings and woodcuts that have survived show a terrific attention to detail. What set Gesner apart from other botanists of that time was that he used every character that he could observe rather than splitting plants into groups on one character alone. This obsession with finding one essential feature was a direct link back to Aristotelian philosophy, and it remained pre-eminent in plant classification until John Locke (1632–1704) introduced the ideas of observations being the route to knowledge and that we are all born knowing nothing. One person who was particularly influenced by Locke was John Ray (1627–1705).

Taxonomy is the science of placing living things in groups (*taxis* is Greek for order or arrangement), and John Ray is the father of modern taxonomy and the second greatest natural historian that England has produced. He was equally comfortable working with plants, animals, and rocks. He was severely troubled by fossils of animals that were no longer living – he referred to them as 'games of nature'. Yet despite his strong religious convictions, Ray laid down the ground rules that have underpinned all taxonomy since.

Ray was the first to define what he meant by a species. This may not seem startling and worthy of note today, because we use the word 'species' in everyday communication and so assume that the

15. Middleton Hall in Staffordshire, where John Ray worked after leaving Cambridge

word has always had a definition and meaning, yet it did not in Ray's time, and in fact even today there are still more than twenty definitions of the word. Species are the building blocks of taxonomy and the currency of biology, so a definition is critically important. Ray's belief was that:

> no surer criterion for determining species has occurred to me than the distinguishing features that perpetuate themselves in propagation from seed. Thus, no matter what variations occur in the individuals or the species, if they spring from the seed of one and the same plant, they are accidental variations and not such as to distinguish a species...; one species never springs from the seed of another nor vice versa.

So a species was a group of similar individuals that can freely interbreed to produce offspring that resemble their parents. This biological definition is still accepted by many and quoted in many exams. There are distinct problems with it when you try to apply it to orchids or to dandelions *inter alia* so a simpler, workable definition is that 'a species is a group of individuals that share a unique set of characters that can be reproduced'.

So Ray recognized that if you wish to maintain the natural variation within a species it should be propagated by seed. This was truly prophetic, as this was 200 years before Mendel's work on genetics (and thus variation) was published and 300 years before the Millennium Seed Bank Project was established at Royal Botanical Gardens Kew to conserve 25% of the world's plant species in the form of seeds. However, Ray's biggest contribution to taxonomy arose from his assertion (following Locke's principles and the practice of Gesner) that if you wanted to arrive at a natural classification, you must use *every* character that you observe and/or measure. You should not ignore anything without good reason. His observations were limited by the resolution of microscopy at that time, yet his work represents a turning point in plant classification. (It should be noted that a *natural* classification to Ray and his contemporaries was one that reflected the *Creator's* plan.)

He dismissed the first division of plants into herbs and trees. He stated that all flowering plants belonged exclusively to one of two large groups that he termed the monocotyledonous plants (monocots) and the dicotyledonous plants (dicots). The monocots he described as those plants in which the seeds have one cotyledon, the floral parts are in multiples of three, the leaves have veins that are parallel, the stems have discrete bundles of vascular tissues scattered through the stem, and the roots are adventitious. On the other hand, dicots he described as those plants in which the seeds have two cotyledons, the floral parts are in multiples of two, four, or five, the leaves have a central vein with lateral and sublateral veins leading to a net-like arrangement, the stems have a ring of vascular tissues around the outer edge of the stem, and the roots are permanent with a persistent tap root and lateral roots. The monocot/dicot split

16. A palm stem showing the bundles of vascular tissue typical of monocots

persisted in plant classification until 1998, and it has to be said that the monocots as defined by Ray still exist. As a further result of Ray using every character that he could see, he also began to put plants into family groups that we still use. The borage family, for example, was one group Ray described, though he did not use the term 'family'.

Ray was by no means the only taxonomist working at this time. He regularly corresponded with Joseph Pitton de Tournefort (1656–1708) at the Jardin des Plantes in Paris. As a taxonomist, Tournefort was a trees and herbs man, but by looking at flower structures he did a great deal of tidying up at the next level up from species in the taxonomic hierarchy, namely the genus (plural genera). This set the scene for Carl Linnaeus (1707–78) to introduce universal Latin binomials, one for each species. During field work, Linnaeus found the long, detailed, descriptive polynomials very cumbersome. He proposed keeping Tournefort's genera but to reduce the rest of the name to a nickname that was hopefully helpful in identifying the plant or remembering it. In 1753, he published *Species Plantarum* which marked the start of legitimate names. He listed every species of which he was aware with one Latin binomial and all the Latin polynomials that had been applied to that plant since Theophrastus and Dioscorides.

It is interesting to note that in one of his four autobiographies, Linnaeus stated that he did not think that *Species Plantarum*, published in 1753, would be his legacy. For most scientists, this would have been enough of a legacy for one man, but Linnaeus was convinced that with his sexual system of classification described in *Systema Naturae*, published in 1735, he had hit upon the perfect natural classification for plants. He grouped them entirely on the number of male and female parts they possessed. His writings were very colourful, and perhaps he set out to shock as much as to enlighten. His sexual system of classification was never universally accepted, though it was useful if you were trying to identify a plant. It did, however, show that the division of plants into trees and herbs was dead and buried.

Linnaeus came up with the idea of classifying plants in this way when he was still in his twenties. He was trying to classify plants from the top down. This is when the characters that you are going to use are determined before you have looked at all the species in front of you. It is very common to find 21st-century undergraduates doing the same thing. John Ray, on the other hand, put all his plants into species with very full descriptions, and when you do this the genera and families almost select themselves; bottom up is better.

The next step forward in the development of our current system of classification was not taken by Linnaeus but by Antoine Laurent de Jussieu, who took the genera of Linnaeus, Tournefort, and others, and placed them in families. The account of this, published in 1789, marks the start of legitimate family names.

As the 19th century began, the plant kingdom had been grouped into species, species into genera, genera into families, families into orders, orders into classes, classes into phyla, and finally phyla into the kingdoms. This hierarchy of taxonomic ranks was a clear and simple way of organizing an ever-increasing amount of data. As plant hunters went about their work bringing plants back to Kew, Edinburgh, and the like, these new species could be inserted into the system. Each species had its own unique position and was given a group at each rank. This made identifying plants easier, either by asking a series of questions, such as it is a monocot or dicot, in order to narrow down the options, or by recognizing the family to which the plant belonged. The botanical family in particular is an exceptionally useful rank, as many of them, such as the orchid, daisy, pea, and celery families, are very easily recognizable.

However, there was revolution brewing as various people began to have doubts about the fixity of species. This was the belief that the world's species were fixed according to the Creator's plan and that they could not change. This went back to Ray's worries about fossils and beyond. Linnaeus had discovered that many plants displayed characters that were a mixture of two others. He often gave these the specific epithet of *hybridus* (for example, *Trifolium hybridus* that still grows on the route taken by Linnaeus when botanizing with his students). How could hybrids arise when all species had been created on Day 1 by God? Linnaeus very neatly got round this problem by suggesting that God created genera and Nature created the species.

As the son of a clergyman, perhaps Linnaeus knew when to keep his head down. However, people like Jean-Baptiste Lamarck (1744–1829) at the Jardin des Plantes were beginning to think openly that species changed in response to their environment. Lamarck is unfairly mocked by some, but Darwin and others give him the credit he deserves. He was one of the people who bravely kicked the evolutionary wasp nest and softened up the scientific world for the theory of evolution so brilliantly encapsulated in the first edition of Charles Darwin's *Origin of Species*.

This book should be compulsory reading for all first-year biology students, and no one should be allowed to begin a degree course until they have read it properly (with their i-pod switched off). The reason for this seemingly draconian idea is that, like Ray, Darwin was a natural historian, equally at home with plants as animals and rocks. His style of writing is still a classic of science interpretation. *Provoke, relate, reveal* is the mantra of conservation education, and that is just what Darwin does. By the time you get to Chapter 13 of *On the Origin of Species*, you are completely drawn into Darwin's line of thought.

Chapter 13 is relevant here, as it is the chapter in which Darwin considers classification. He believed that if his idea of descent with modification was correct, then this would go a long way to explaining the fact that we had classified biology into this hierarchical system of nested ranks. The only diagram in *On the Origin of Species* is a branching tree – the era of phylogenies had begun. Darwin believed that species could be placed in genera because they shared, at some point in the past, a common ancestor – *commonality of descent* – and that the further you had to go back to find that common ancestor, the further up the hierarchy of ranks you went – *propinquity of descent*. The system of classification did not prove evolution had happened, but the system of evolution proposed by Darwin explained why we could classify as we did. At a stroke, the meaning of *natural classification* had changed. No longer were taxonomists trying to reveal the Creator's plan; now they were replaying evolutionary history. This in itself was one of the great objections to Darwin's theory, because whereas the Creator was believed to have had a plan, Evolution has no plan, and that frightened people.

As the 19th century rolled on, the number of plants in herbaria such as at Kew grew, and the classification was adjusted and refined. Between 1862 and 1883, George Bentham and Sir Joseph Hooker published their classification of plants. In the flowering plants, they recognized monocots and dicots. Within the dicots, the first divisions were into three big groups: plants whose flowers had petals that are free from each other, plants where the petals are fused into a tube, and those where there appear to be no petals. Within the monocots, the divisions were based on an eclectic range of characters such as whether the flowers were colourful or brown and membranous, whether the seeds were tiny, and so on. Bentham and Hooker's system was very useful for identification purposes, but they made no attempt to reveal evolution despite Hooker being a major correspondent with Darwin. Perhaps it was just too controversial.

The publication of *On the Origin of Species* opened the door for the creation of phylogenies. German polymath Ernst Haekel (1834–1919) is often given the credit for coining the term 'phylogeny'. A phylogeny is an evolutionary classification and is drawn as a branching tree with each end branch representing a group. It is hoped that one day we shall be able to create a phylogeny for all flowering plants and there will be about 350,000 terminal branches, each one representing a species. Haekel's trees were based on the morphology that he and others could see with increasingly sophisticated microscopes, so essentially they were using similar evidence to that available to Ray and his colleagues.

In the 20th century, a new possibility arose. With the discovery of the role of chromosomes and then the structure of DNA in 1952, followed by the deciphering of the genetic code, people began to wonder if this was the essence about which Aristotle had written. Would it be possible to find a stretch of DNA that is unique to every member of a species? Could we use the difference in the sequence of bases in a stretch of DNA taken from two different organisms to determine their propinquity of descent? These are in fact two different things. The first is identification, the second is classification. Before either of these questions could be answered, another technical innovation rocked the taxonomic boat, and this was the invention of the electron microscope that revealed a box full of new characters to be considered.

One of the most beautiful plant structures revealed by electron microscopists is the patterning on the outside of pollen grains. This is the sporopollenin that has featured so much through this book. The photographs taken with electron microscopes revealed not only patterns that can be unique to a particular species, but also that pollen grains have a number of apertures through which the pollen tube grows after the stigmatic surface has given it water. The number of apertures varies from one to several, but it is usually one or three. This was a hugely significant character because it helped to solve a problem that had become increasingly awkward.

Along with the addition of structures revealed by electron microscopes and the addition of DNA sequences (also known as molecular data), there had been one very major shift in the rules. This was and is the principle of monophyly. A monophyletic group is one that contains all of the descendants of a common ancestor. To put it another way, all the taxa in a monophyletic group share a common ancestor, and the group must contain all of the descendants of that common ancestor. Monophyly can be applied at every rank. So all living organisms form a monophyletic group because life has evolved once. Plants (as defined in this book) are a monophyletic group, as are land plants, seed plants, flowering plants, monocots, orchids, and so on. None of the new evidence of characters has shaken the monophyletic status of the monocots as described by John Ray over 300 years ago. However, the dicots have not fared so well.

When Darwin began to think about how particular lineages might be related, he came up against a problem. Although seed plants (gymnosperms and angiosperms) appeared to be a monophyletic group, that is to say that seeds have evolved just once, what did the common ancestor look like, and what was the origin of the angiosperms? He described this to Hooker as 'an abominable mystery', and mystery it remains. One way of trying to reveal the truth is to find out what the first flowering plants looked like. Before the molecular data and electron micrographs became available, botanists were beginning to think that magnolias represented an early group because the oldest fossils of flowering plants looked very similar to modern magnolias, and because magnolias are pollinated by beetles, which were present before bees and were around when these fossil flowers were living. In addition to the magnolias, the waterlilies, peppers, bays, and several other groups were being examined. All of them were dicots. The pollen evidence then threw a large spanner in the works. It turned out that all these groups which were suspected of being in some way odd all had pollen with just one opening, like the monocots. All the dicots which appeared to be in the clear had pollen with three apertures.

When the molecular data was thrown into the mix with all the macro-morphology and micro-morphology, it appeared that the flowering plants are made up of two very big monophyletic groups, the monocots and the true dicots (officially the eudicots), and lots of little groups. This seemingly heretical suggestion was first aired in 1993, and it was considered so radical that half of the contributing scientists withdrew their names from the paper. By the time the second version was published in 1998, no one was in any doubt that the monocots and eudicots were here to stay, but the others needed to be sorted out. It is now thought that the 352,000 species of flowering plants should be grouped into 13,500 genera in 462 families in 41 orders (or thereabouts!).

The 1998 paper was a landmark in plant taxonomy. Firstly, it had a huge author list because it was a multinational collaboration, nominally led by Mark Chase from the Royal Botanic Gardens Kew. However, rather than being cited as Chase *et al.*, the paper's authorship is APG – the acronym for the Angiosperm Phylogeny Group. Secondly, the paper was rejected by both *Nature* and *Science*, leaving the Missouri Botanic Garden to publish it in their *Annals*. The biggest paper in plant taxonomy for 150 years got more coverage in the UK press than in the elite scientific journals, clearly illustrating the depth to which taxonomy had sunk in the eyes of the scientific establishment. Thirdly, the APG said they did not know where to place certain groups. This was unheard of. All previous classifications were complete, with every taxa assigned to a place at every rank. Since APG I, there have been APG II (2002) and APG III (2009). The detail is becoming clearer. The lowest branch on the flowering plant phylogeny is *Amborella* from New Caledonia. Being the lowest branch means that this species had a common ancestor with all other species of flowering plants longer than any other group of flowering plant. They are often referred to as a relic group. Next is the waterlilies and a few allies; next the Austrobaileyales order, a group of families including most famously star anise, which for several years was the origin of tamiflu, a drug that was going to treat the pandemic of swine flu in 2009. After that, the picture is murky. There appears to be a large group consisting of the magnolias, black peppers, bay, and avocados, but we do not know how this relates to the monocots and dicots or a couple of oddities, *Ceratophyllum* and *Chloranthus*. Only time will tell, but as yet Darwin's abominable mystery is still a mystery.

After all this work to classify the world's plants, one might ask 'so what?'. Is this just a job-creation scheme for vegetation trainspotters? Is a subject with an acknowledged subjective and philosophical component really a science? The latter is easily answered. Taxonomy is trying to make order out of random evolution. Species are constantly changing. There is no fundamental difference between a species and a variety of that species (Darwin, 1859). Taxonomy is therefore like nailing

17. *Magnolia* **flowers appear to have existed for at least 115 million years**

jelly to the wall – messy, but worthwhile for the following reasons.

Firstly, a phylogenetic classification enables us to make objective decisions in conservation where inadequate resources have to be allocated to the places of greatest need. Take, for example, two similar areas of vegetation each containing 8,000 species but you can only conserve one area. If one contained plants from 100 families in 20 orders, and the other plants from 50 families in 10 orders, then the former probably contains more of the evolutionary tree than the latter and so is more valuable.

Secondly, as indicated earlier, some parts of the evolutionary tree are richer in medically efficacious materials than others. The potato family (the Solanaceae) has given us atropine and hyoscyamine, among others. When the power of galanthamine to treat Alzheimer's disease was discovered, it was only known from snowdrops. It was also known that snowdrops could not grow fast enough to supply the huge demand. An alternative supplier was quickly found in the Amaryllidaceae, in the shape of daffodils. Likewise, although taxol for the treatment of ovarian cancer was originally extracted from the Pacific yew (*Taxus brevifolia*), it is not a sustainable supply, and an alternative supply of the raw material has been found and commercially extracted from English yew (*Taxus baccata*). In the past year, harringtonine has been licensed for the treatment of leukaemia. Harringtonine was first found in *Cephalotaxus*, a close relative of yew trees.

But drugs are not the only things we get from plants, so what have plants ever done for us?

Chapter 6
What have plants ever done for us?

It has been said that there are three broad reasons for conserving plants: the 3Ss or the 3Ps or the 4Es. These are sensible/selfish/spiritual or pragmatic/practical/philosophical or ecological/economic/ethical/aesthetic (well almost). There is little doubt that plants (and biology in general) improve the quality of our lives. They are beautiful, intriguing, and their cultivation is regarded by many people to be therapeutic. The role of plants in art and the art of landscape design cannot be ignored, but these are personal aspects of plants that fall under the spiritual/philosophical/aesthetic banner.

This leaves two broad groups of reasons why plants are important. On the one hand, there are the big ecological services that are normally presented under four headings: provisioning (e.g. food or fresh water), regulating (e.g. carbon sequestration), supporting (e.g. providing pollinators), and finally cultural (e.g. hiking). To a degree, these services are not species specific, so it can be argued that it does not matter which tree species is growing and absorbing carbon so long as a tree is growing. It has been calculated that the value of theses ecosystem services is approximately $33,000,000 million per annum. This is a strange, sterile calculation because if you had $33,000,000 million in you wallet, where would you go to buy these ecosystem services?

Furthermore, the global gross national product for the planet each year is just $18,000,000 million, so even if it was for sale, we could not afford it. If something is not available for purchase, is it priceless or valueless? Perhaps this accounts for the cavalier manner in which biology has been treated in the past few centuries.

The other reasons for looking after our biological heritage are species specific. These are those commodities and resources that we extract or otherwise derive directly from the herbage. Included here are plants like the English yew, which supplies us with Bacattin III used in the synthesis of taxotere for the treatment of breast cancer.

Asking what have plants done for us may have become a question that needed to be answered only in recent decades as some of us become increasingly removed from the plants that we exploit. On one simple level, plants have done, and will continue to do, everything for us because of their (as yet) unique ability to use the Sun's energy to make an awe-inspiring range of chemicals, some of which all the Queen's chemists cannot synthesize.

On another, very biological, level, plants create the habitat that we are currently occupying with at least 1.5 million other species. Yet one of the things that sets *Homo sapiens* apart from other species is its ability to increase the carrying capacity of its habitat. It has been suggested that in the long term, sustainable agriculture can only support 1,500 million people, though in the medium term it will be required to support a predicted maximum 9,500 million by 2050. Without agriculture, the planet will support approximately 30 million hunter-gatherers. This is a sobering (or just silly) figure since there is nowhere for the other 6,470 million people to go. In order to feed the current population, we are cultivating one-quarter of the total land area, and we are currently consuming, in one way or another, 40% of all photosynthetic activity. It is of great concern that if we have to expand the area under cultivation, then many other species are going to be evicted from their current habitats.

The origin of agriculture is by no means clear. The traditional view that persisted until the end of the 20th century was that about 8,000 years ago (YBP) grasses were domesticated in the region between the River Tigris and the River Euphrates, henceforth known as the Fertile Crescent. It is believed that as many as eight crops were grown here: einkorn wheat, emmer wheat, barley, lentils, peas, flax, bitter vetch, and chickpeas, with beans appearing later. At about the same time, rice was domesticated in China and potatoes bred in South America. After that time, more independent centres of domestication emerged, including the breeding of maize in Central America from teosinte. What could have stimulated this simultaneous, but uncoordinated, adoption

18. Sweetcorn is now one of the three major crops

of an agrarian way of life? The climatic aberration known as the Younger Dryas Event was traditionally given the credit for being the catalyst. This simple view is now being challenged.

It is now thought that the end of the last ice age, c. 14,500 YBP, was marked, in the Middle East and probably elsewhere, by an increase in temperature closer to present values. The temperature then promptly returned to ice-age values about c. 13,000 YBP. This is known as the Bølling-Allerød Interval. This was followed by the cold but dry Younger Dryas Event, c. 12,900 to c. 11,600 YBP, at the end of which the temperature rose to far above that of the ice age and the glaciers finally retreated. Following the end of the Younger Dryas, the climate became unfamiliarly constant from year to year, and so humans stopped moving around in groups of between 15 and 50 and they started to live in settlements that were increasingly made of stone. These are known as the Natufian people and they lived in what is now Israel, Jordan, Syria, and the Lebanon. Olives, pistachios, wheat, and barley began growing in this area as the temperature rose. These plants were pre-adapted to these new conditions.

There are now about 60 sites that were populated by the Natufians, but there is very little direct evidence about which plants were being used. One of these sites is at Abu Hureyra in the Euphrates Valley in Syria. There are two sets of remains here from before and after the Younger Dryas Event. It is thought by some that it was the cold, dry weather that forced the people in these areas to cultivate grains, which had become much rarer in the wild. Nine plump rye seeds have been found at this site. Some researchers say this is hardly firm foundations for an entire theory, not least because rye has not been found elsewhere and does not make another appearance for a few millennia. It may be that the 50% increase in carbon dioxide levels at the end of the Younger Dryas Event helped by making agriculture photosynthetically supportable.

There is some agreement about how *Homo sapiens* brought plants under his control. We know that hunter-gatherers ate fruits, roots, seeds, and nuts from wild plants. Included here are grains from grasses. In 2009, evidence was presented that appears to show that grains, and in particular sorghum, were being ground up using stone tools in Mozambique 105,000 YBP. This evidence is supported by the presence of amyloplasts in the tools. Amyloplasts are the organelles where plants make and store starch. The pattern of the deposition of the starch and the size of the amyloplasts is often species specific. Furthermore, cultivated varieties often have larger amyloplasts. The amyloplasts of wild chillies are 0.006 millimetres long, while cultivated varieties are 0.02 millimetres long. Amyloplasts are really useful because they are resilient to decay and are found not only in early kitchen utensils but also in sediments. They have now been used to date the domestication of squashes, manioc, and chilli peppers in America.

On the south-west shore of Galilee is the Ohalo II site, which was inhabited 23,000 YBP. More than 90,000 plant remnants have been found here from *inter alia* acorns, pistachios, wild olive, and lots of wheat and barley, but nothing that looks like a cultivated variety (cultivar), and there is no evidence for the grinding up of cereals.

It is important to this story to consider what is meant by cultivation and domestication. The former is simply the deliberate growing or protection of a wild plant. Growing begins with a propagule such as a seed or cutting, while protection is where a particular plant is preserved and nurtured because it has something that is wanted. It is often difficult to prove cultivation of wild plants. Domestication, on the other hand, is visible. There are many definitions for domestication, but they all include an element of permanent physical and genetic alteration to improve the plant and make it more suitable for the needs of humans.

These improvements are known as domestication syndromes. Some of the characters are obviously better for the farmer. These include compact growth, making the plant more resilient, simultaneous ripening allowing one harvest, loss of seasonal flowering and/or germination permitting sowing at various times of year, and increase in the size of the grain with a thinner seed coat for increased palatability and easier processing such as grinding and milling. Enhanced flavour and nutritional value are characters that the consumer would like to see, and when agriculture began the farmer and consumer were the same people (as they still are in many parts of the world).

For some crops, there are very specific obstacles to overcome during domestication. In figs, this is the requirement for the flowers to be pollinated by a specific species of wasp before the 'fruit' that we eat can develop properly. The flowers are enclosed in the syconium (the fleshy part of the fig that we eat and which is, strictly speaking, not a true fruit). The figs that we buy in greengrocers are a mutant variety that will develop its fleshy syconium without the need for the pollination of its flowers. These parthenocarpic varieties are ancient. Remains of figs excavated at Gilgall, a neolithic village in Jordan, show extraordinary detail of one of these ancient parthenocarpic fruits that has been accurately dated at 11,400 years old.

One character that can frustrate farmers is the propensity for plants to drop their seeds when they are ripe. The farmer wants the plant to hang on to its seeds and fruit until he is ready to harvest them. This means that seeds from domesticated varieties do not have a smooth abscission wound but a jagged scar or tear. This is a very useful trait for archaeobotanists because it can be found on the remains of very old seeds. Such seeds have been found in the remains of the Nevali Cori settlement in Turkey. Seeds found here were harvested 10,500 YBP, and they have a jagged abscission layer indicating that they did not fall naturally but were torn off during the harvest. This is the earliest direct evidence for the domestication of plants.

However, it should not be assumed that the transition from collecting wild grains to growing fields of selected and bred varieties happened cleanly one year. There is good evidence, also from Nevali Cori, that wild and domesticated plants were grown side by side, perhaps because the farmer could not prevent some of the wild seeds from falling before harvest thus enabling them to persist from year to year in the soil seed bank. It is now widely accepted that domestication was preceded by many years of cultivation and so a possible scenario for the transition from hunting and gathering to farming may have involved four stages. The first involved the hunter-gatherers moving around in small groups, their movements dictated by availability of food and by the weather, which could be very inconsistent. These wanderers would inhabit caves where possible, and in these is found debris from their everyday lives. More than 1,000 fibres of flax have been found in caves in the Caucasus Mountains. These sites were active 36,000 YBP. These fibres appear to have been dyed, even though dying flax is very difficult.

It is known that *Homo neanderthalensis* and subsequently *Homo sapiens* had learned that fire could be used to great advantage. Selective burning of small areas provoked the regrowth of soft, palatable vegetation. Lignified tissue is almost impossible to digest, but cellulose poses less of a problem to some grazing animals. This means that the regenerating burned area would attract animals that could be picked off easily by men with spears waiting in the unburned bushes.

The second stage in the adoption of an agrarian way of live would have involved a combination of hunting, gathering, and cultivation of wild plants. In this way, the skill of farming would have been learned and understood.

Having learned how to grow plants, the early farmers were able to turn their attention to the selection of better plants by looking for the domestication syndromes described earlier. Thus, the third stage of the process involved a decreasing reliance on hunting and gathering and more dependence on the produce of fields. These fields would have contained an increasing proportion of domesticated varieties of both plants and animals. Evidence from the Ohalo II site indicates that 10,000 YBP, just 10% of cultivated plants exhibited domestication syndrome. This proportion had risen to 36% by 8,500 YBP and to 64% by 7,500 YBP. It is also becoming clear that different crops were domesticated at different rates, that those rates were not constant but punctuated with bursts of activity, and that different crops were altered in different ways.

It is now believed that although the Chinese began consuming rice 12,000 YBP, it was in fact millet that they began to domesticate first, around 10,000 YBP. In the Chengdu Plain in western China, millet was the staple diet 4,000 YBP and was used to make flour, porridge, and beer. Evidence from China shows that the domestication of rice was a slow process. Around 24,000 pieces of plant have been examined from just one site, including 2,600 rice spikelets. While rice domestication may have started 10,000 YBP, by 6,900 YBP only 8% of cultivated plants were rice and just 27.4% of this was domesticated. By 6,600 YBP, rice accounted for 24% of all cultivated plants but only 38.8% was domesticated. As cultivation spread, so did the extent to which the environment was altered to accommodate fields. In the lower Yangtze region, alder woodland was cleared c. 7,700 YBP and the water levels managed for 150 years by the use of bunds until a disastrous flood c. 7,500 YBP.

The final stage in the evolution of *Homo sapiens* into an agriculture-dependent species involved an increase in the area of land under cultivation and a spreading of this way of life. It is now believed that although squash, peanuts, and manioc were growing in the Andes 10,000 YBP, they were originally domesticated in the lowland tropical forests. The evidence for this comes from the study of the genomes of the wild and cultivated forms. There is also evidence of earthworks in western Amazonia that is the proposed centre for the domestication of not only squash, peanuts, and manioc, but also chilli, beans, rubber, tobacco, and cocoa.

It is believed that maize was being grown in the Andes between 7,000 and 5,000 YBP. However, we do not know how different these plants were from the original teosinte from which modern maize has been bred. While modern wheat and barley both appear to have been bred in more than one place, with more than one original hybrid or selection, modern maize all stems from one domestication event, at some time between 9,000 and 6,000 YBP. The transformation is eye-watering. For example, the maximum number of kernels on teosinte is about 12, whereas a cob of corn will now contain up to 500 or more as a result of selecting plants with more and more kernels and growing and crossing only those plants.

It is now believed that there are more than 24 regions where cultivation began independently. In 13 of these areas, grains were the major crops. As a rule of thumb, increasing seed size and non-shattering seed heads were the first two domestication syndromes to be worked on. For pepo squashes, it was not only larger seeds that were selected but also thicker stems and thinner rinds.

Following the emergence of these two dozen centres of domestication, agriculture began to spread all over the world as people moved about. Sometimes they took their plants with them. For example, the African gourd (*Lagenaria siceraria*) has been found in 10,000-year-old settlements in America. It is believed that it was taken there by Palaeoindians perhaps via the coast of Asia rather than across the Atlantic. There is good evidence that the early farmers in Europe were migrants who taught the indigenous population how to farm. It is not clear yet whence they came but by bringing cultivated varieties of plants with them, and thus taking the cultivars away for their parent species, they were inadvertently speeding up the process of domestication by preventing genetic dilution of their selections by back-crossing with the parent stock. In effect, they were separating the gene pools.

While food plants were being domesticated, other natural resources were being extracted from plants. Rubber and fibres have already been mentioned, but trees were exploited for their timber and their leaves in the construction of dwellings and the heating of those homes. However, another major area of use was in medicines.

It is difficult to the say when and how early hominids began to appreciate the healing power of plants. Again, it is debris in caves that gives us some clues. In caves in northern Iraq, pollen grains are found in large numbers, and it is assumed these are from the plants that were exploited by the cave's inhabitants, be they *Homo neanderthalensis*, *Homo sapiens*, or the recently discovered hybrids. About 90% of the pollen grains have been identified as being from plant species still used in this part of the world for their medicinal properties. How these properties were discovered is a subject for speculation, but there are several possibilities.

Serendipity cannot be ruled out, as simple chance observation; certainly, this was how people would have discovered which plants were toxic. Observing animals may have made people inquisitive about plants. Echinacea, willows, and tea tree are eaten by mongoose, horses, and koala respectively. All have proven beneficial effects. Various strange ideas have made a small contribution to our knowledge. In pole position here must be the Doctrine of Signatures, based on the idea that if a part of a plant resembles a part of the body, then the plant should be used to treat ailments of that part of the body. Only *Aristolochia* used in encouraging

19. Birthwort has been used for centuries to induce childbirth

childbirth, and *Podophyllum* used to treat testicular cancer, can be claimed as pseudo-proof for this charmingly dangerous idea.

Traditional Chinese medicine is probably the most widely used form of herbal medicine. The oldest printed herbal is the *Pen Tsao* printed in China 4,000 years ago. Many of the infusions and tinctures recommended in this work are still used in China today and still work. One of the problems for science is discovering the active ingredients in these preparations because they often contain more than one plant, and so the active ingredient may be a combination of several ingredients working together. A very extensive programme is currently underway in China that aims to uncover the active ingredients in all 6,000 species that are used in traditional Chinese medicine.

The Chinese have used *Ephedra* for more than three millennia in the control of bronchial asthma, and it is still a common ingredient in children's cough medicine. It is also used by European anaesthetists to prevent the patient coughing during the operation, something that could be bad news. They also use it as the patient comes round as it mimics the effect of adrenalin, thereby making them feel re-invigorated and ready to go home. It should be noted that the International Olympic Committee and others regard ephedrine and its derivatives as a performance-enhancing substance. The use of *Artemisia annua* in the treatment and prevention of malaria has been derived straight from Chinese herbal medicine, though the mode of action is still poorly understood. For many years, extract of *Indigofera* has been used to treat 'bad blood', and it is now used in the treatment of leukaemia in the West.

The use of *Ginkgo biloba* in China to treat circulatory problems, especially to the brain, remains to be proved, but the author's mother found it was terrific for improving circulation to her hands and feet when nothing prescribed by the GP had worked. However, her heart is being kept going by the extraordinary ability of digitoxin to regulate and strengthen heart beat. The original use of *Digitalis purpurea*, and more recently *Digitalis lanata* digoxin and then digitoxin, was first proved by a Dr Withering working in the village of Edgebaston in the 19th century, when Birmingham was smaller and cricket had not been created. He had come across the idea from a Shropshire Romany tincture from 20 herbs that was claimed to be efficacious against a wide range of ailments. By process of elimination, the foxglove was found to be the hero of the hour.

With the advent of the scientific method, more rigorous testing was carried out on the claims made about the therapeutic value of plant extracts. The problem was finding willing subjects upon whom ideas could be tested. In the 16th century, an early form of the parole board in prisons enabled prisoners to be released early if they allowed physicians to observe the effects of various treatments on the inmate. Early release was guaranteed – but

20. The flowers of English foxgloves

often only in a coffin. The toxic properties of hellebores were discovered in this way.

Clinical trials are now more tightly controlled, and the search for new treatments more systematic. However, the value of local knowledge is still inestimable. The discovery of taxol is a direct

102

result of the native North Americans sharing their traditional medicinal knowledge with the US National Cancer Institute. Having extracted the taxol from the bark of *Taxus brevifolia* and demonstrated its potential against ovarian cancer in mice, the research was taken up by a commercial pharmaceutical company. Their first problem was getting hold of sufficient taxol to carry out the research. It was not possible to synthesize the stuff, and so the only option was to harvest it from the bark of the trees. While this helped the company, and ultimately the patient, it did nothing for the trees. This was never going to be a sustainable supply of taxol.

In Europe, where researchers had no access to the trees of *Taxus brevifolia* (because they belong to the American State), an alternative production system had to be found. Eventually, the leaves of English yew (*Taxus bacatta*) was found to contain Bacattin III, which can be changed into medically efficacious taxol and subsequently taxotere for the treatment of breast cancer. In addition to showing that new, modern, clinically proven treatments are still being derived from plants, this story raises three other important issues. Firstly, we should look after all species. Even if we do not exploit them at present, we may in the future. Secondly, the sustainable extraction of biological resources is very important. There is no biological credit card. That being said, we are using up our reserves at an impressive rate. It has been calculated that in a single year, we release carbon dioxide from fossil fuels that took plants three million years to fix by photosynthesis. Thirdly, this raises the question of ownership. The US took the view that taxol is theirs and no-one else can exploit it. (In the end, it transpired that they have rights to the ownership of *Taxus brevifolia* but not the patent on taxol.) These three issues are the three headings under which the Convention on Biological Diversity derives its framework.

Living within our biological means is an important issue that will not go away. Photosynthesis is the only way of providing us with our daily food and so much else. It has been possible to increase the output per acre for many of our staple foods. The most recent green revolution was led by Norman Bourlaug and others. In 1970, Bourlaug was awarded the Nobel Peace Prize in recognition of the impact that he had had on crop yields in those parts of the world where malnutrition was a way of life. Between 1966 and 1968, the yield of wheat in India rose from 12 million tonnes to 17 million tonnes. A plant breeding programme for rice resulted in a similar increase in yields. Sadly, the results in Asia were never transferred to African agriculture, perhaps as a result of the lack of infrastructure such as roads, irrigation, and seed production. Bourlaug's mantra of 'impact on farmers' fields, not learned publications, is the measure by which we will judge the value of our work' is not quite in line with current science research funding in the UK.

It is estimated that between 1960 and 2000, the proportion of the world's population who felt hungry for at least a part of the year fell from 60% to 14%, though this is still almost 1,000 million people. The next green revolution will have to increase yields further. This may be possible if waste and loss to pests and diseases can be reduced. Growing drought-tolerant crops such as sorghum might be another option. One exciting possibility is to alter the photosynthetic machinery in rice to make it more like maize. (If you are motivated to learn more about this, rice is C3 and maize C4, which means that rice fixes carbon dioxide into a molecule with three carbon atoms whereas maize fixes carbon dioxide into a molecule with four carbon atoms.) This would require some clever modification of the genetics of the rice. Such a modification appears to have happened 45 times in the course of evolution, but this is different, and some people are frightened of further genetic modification of our food plants.

The risks of genetically modified crops fall into two broad categories: human and animal protection and ecological protection. There are already procedures to ensure that food sold in the UK is safe to eat. Ecological protection is more difficult but can be tested and assessed. The three major risks may be summarized thus. Will the genetically modified variety escape from its field? Will it hybridize with native species? Will it cross-pollinate non-genetically modified crops of the same crop plant? If the answer to all of these is no, then perhaps the risks are acceptable. Ultimately, it will be the lack of an alternative that will drive people to embrace genetically modified crops and the food that results.

The green revolution of the 1960s and beyond relied on high inputs of water and fertilizer. We know that phosphorus supplies may peter out in the middle of the 21st century, closely followed by the supplies of inorganic nitrogen. Fresh water supplies are not infinite. This is a particularly relevant problem in the drive to produce bioethanol and biodiesel for our vehicles. It has been calculated that while oil extraction and refining may use up to 190 litres of water per kilowatt hour, and nuclear power up to 950 litres of water per kilowatt hour, the production of corn ethanal requires up to 8,670,000 litres of water per kilowatt hour, and soybean biodiesel production up to 27,900,000 litres of water per kilowatt hour. It appears that biofuels will make us all very thirsty.

Chapter 7

Looking after the plants that support us

On 5 June 1992 in Rio de Janeiro, at the first Earth Summit, the Convention on Biological Diversity (CBD) was opened. To date, the vast majority of nation states have signed and ratified the convention that aims to conserve the components of biodiversity, to ensure its sustainable exploitation and to facilitate the equitable sharing of the benefits of this exploitation. As a result of this landmark agreement, there was a huge increase in conservation activity *sensu lato*. Each country drew up an action plan and took responsibility for the implementation of the CBD in their region. That this happened is supported by the fact that 90% of money raised for conservation is spent in that money's country of origin. An example of this is the Millennium Seed Bank Project at the Royal Botanic Gardens Kew that had as its first target the collection and safe storage of almost all the around 1,500 species of plant native to the UK. In addition to the increase in practical activities, there was an explosion of scientific papers mentioning biodiversity in their title.

By the end of the 20th century, it became apparent to many botanists, especially those working in botanic gardens around the world, that although there was a great deal of activity, there was no way of knowing if the aspirations of the CBD were being met. Following a meeting in April 2000, the Gran Canaria declaration was published calling for a Global Strategy for Plant Conservation

(GSPC) and suggesting more than a dozen targets to be hit by 2010. This idea was discussed by the countries that had ratified the CBD, and in September 2002 the GSPC was adopted, with 16 targets to be hit by 2010. This was perhaps a surprise, as never before had a target-orientated strategy been proposed for a major group of organisms. The rest of the conservation community is very interested to see if this approach works because it may be a model for all conservation work in the future. By 2010, it was clear that some of the targets had been hit, some were going to be missed but not by much, and that a couple were going to be missed badly. As a result, GSPC 2 was drawn up during 2010. This was clearly based on the successes and failures of GSPC 1; the progress of plant conservation was much better informed in 2010 than it was ten years earlier.

The GSPC is a wonderful framework, or list, of what we need to achieve if we are going to realize the aspirations of the CBD in relation to plants. The 16 targets cover all areas of activity related to plant conservation. It enables different organizations to make a contribution commensurate with their resources and missions; it is acknowledged that neither one organization nor one person is going to save all of the world's plant species from extinction. In order both to understand the scale of the task to halt and then reverse the decline in plant species and to appreciate that this is not an insurmountable problem, it is helpful to go through the targets of the GSPC.

Target one is to draw up an online list of all the known plant species in the world. This task is being coordinated by the Royal Botanic Gardens Kew and the Missouri Botanical Garden which between them have the most comprehensive herbarium collections in the world. The problem of compiling this list is that there are two ways of cataloguing species. One is by monographs that survey all the species in a genus (or part of a genus, if the genus is large). This work is generally carried out by one person who takes a global view of the diversity within the species. The other type of catalogue is the national inventory found in a flora. While national pride is often a potent driving force behind conservation work, it is well known that local botanists will often seek to inflate the number of species in their country by raising to the rank of species small variations that are just the normal range of diversity found within a species. This innocent chauvinism means that you cannot simply add up all the species in floras. Nor can you just make a list of all the legitimately described species, because most species have been named more than once and so there are synonyms to sort out.

The work to compile the world checklist, as a first stage in the creation of a world flora, has been organized by botanical family and this has thrown up some anomalies. There appears to be no pattern to the areas of ignorance. Large families are no more difficult than small families. Families that have a long history of cultivation, for either economic reasons or aesthetic pleasure, are no more likely to be well understood than those that are currently of botanical interest only. This work is therefore showing where there is a lack of expertise – this is referred to as the 'taxonomic impediment' in Europe and as the 'taxonomic abyss' in Australia!

Having put together the list of species, we need to know which are struggling to survive and this is target two. There is a long tradition of countries and regions compiling 'red lists' of threatened species under the aegis of the International Union for the Conservation of Nature (IUCN). These have been compiled on a national basis for many countries, but bizarrely these suffer from the same inflation seen in floras. It appears that long national red lists are seen as a way of levering more money out of funding bodies. National lists also lead to species that are very common in one country but rare next door appearing to be threatened; false positives are common.

There is a global list that includes an assessment of 47,677 species across all taxonomic groups. This list currently (2009) shows that

17,291 species, or 36%, are threatened with extinction in the next 50 years. The statistics for plants are rather worse, with 71% of the 12,151 species assessed being either extinct, extinct in the wild, critically endangered, endangered, or vulnerable, with just 12% being categorized as least concern. These figures reveal one of the problems connected with data-gathering, that of definitions. The IUCN categories are clearly defined and are the gold standard, but they take a great deal of time to compile, time we may not have.

It has become apparent that an heuristic approach is required in the first instance. Such a system has been devised, the RAMAS Rapid List, that reduces to just three categories: likely threatened, likely not threatened, and likely data-deficient. Or to put it another way: threatened, OK, or don't know. This allows workers to input data quickly, without the need for lengthy calculations and comprehensive details. In addition to providing a more broad-based approximation of the problem, this will show where the data deficiencies are and thus where more resources are needed. It is now expected that all new monographs will include an assessment of the threatened status of all the species described. For some genera, such as magnolias and maples, genera-based assessments have been carried out with full monographic revisions. Other estimates of the number of threatened species exist, and generally the numbers are between 22% and 47%. It is reasonable to assume, therefore, that the general 36% figure is not unreasonable, and so if there are 352,828 species of flowering plants, then 127,018 are threatened with extinction in the next 50 years.

It is already clear that there is a great deal of work being done, and more is required if just targets one and two are to be hit. Sound scientific research in all aspects of conservation biology is required, and the execution and dissemination of this research is target 3. While peer-reviewed journals are always going to have their place, it is equally true that there is great deal of work going unreported and yet which would be very helpful to conservation workers, especially those in the field of developing countries. Since 2002, a number of web-based resources have emerged where evidence-based conservation can be posted. A recurring experience of conservation projects is that each one is slightly different. Each species recovery project is different and each habitat restoration project is different because the assemblage of species is different from site to site. While the CBD celebrates and protects biological diversity, that same diversity makes generalizations difficult and makes rules loose.

So the first three targets are about understanding and documentation without which the next seven targets would be unattainable. Target four is to protect at least 15% of the world's main ecological regions. It is currently estimated that 11.5% of the land area of the world is under some form of protection. In China, this figure is 14.7%. The advantage of conserving ecological regions is that you conserve the ecosystem services described in the previous chapter. Furthermore, it is often claimed that if you have a healthy, stable, diverse community of plants then the animals, fungi, and other organisms will be there too. It is for this reason that the diversity of plant species is often used as a surrogate measure for all of biodiversity. This is not unreasonable, as most ecosystems are described by the plants that dominate them. It is furthermore claimed that by protecting large areas, you are protecting not only the species that you have recorded, you are also looking after the unknowns. This is particularly true for insects, of which perhaps only 10% have been described. It is not as important for plants, as it is believed that 90% of all plant species have been named at least once.

While the protection of ecosystem services is undeniably important, it is equally true that plant diversity in general, and individual species distributions in particular, are not evenly spread throughout the world. This problem is addressed by target five, whereby 75% of the important plant areas are to be protected. The objective here is to try to avoid gaps in the coverage of plants at all latitudes and types of habitat. It is well known that, generally speaking, the number of plant species per unit of area falls as you get further from the equator, but this does not mean that the temperate regions are less important and valuable than the tropical regions. It has also been found that deciding where to place your protected areas can require some very complicated analysis of data. A study in Madagascar attempted to optimize the allocation of protected areas for 2,138 species of plants and animals. They were only permitted to protect 10% of the island, but each group of plant and animal required a slightly different 10%. In the Philippines, another study showed that while many plants were adequately protected in nature reserves, the threatened palm trees grew mostly outside these areas.

While protecting areas of 'natural' vegetation will always be seen as a good idea, it should not be forgotten that 25% of the land on

21. Palmer's Leys is a field in Oxfordshire that has been restored to wild flower meadow in just three years. Since 1950, 96% of this type of meadow in the UK has been ploughed up

Earth is under some type of production regime. This does not have to be an intensive agricultural system with high inputs of fertilizers, pesticides, and perhaps water. It can be a hill farm in Wales or a cork oak woodland in Iberia. The latter is a very good example of a commercially supportable production system that has an associated flora that puts it in the top 20 regions of the world for plant diversity. Target six aims to have 75% of production lands managed with the conservation of plant diversity as one of the aims. This is more easily done for woodland than for a farmer's field, and yet stone curlews are often seen nesting in intensive sugarbeet fields in East Anglia. Worldwide, as much as 60% of forestry land has the conservation of the biodiversity written into the management goals for the area.

So these three targets are concerned with large-scale, vegetation-based conservation. However, much of conservation is at the level of species, not least because people can see that they are making a difference if they champion and protect a species. There is a risk that species-based conservation is not coordinated and the species chosen for conservation are not the most deserving but perhaps the most iconic. It is, for example, easier to fund orchid conservation than the protection of a species of moss. That being said, a great deal of conservation work is carried out at the level of species.

For many years, the ultimate goal was to carry out all species conservation *in situ*, and if the species was declining in its habitat then you would grow some more and reintroduce it. This failed far more than it succeeded because, unless the threat or reason for the decline is removed, then the new plants will go the same way as their predecessors. The only threat that could be removed easily was over-collecting by humans. This led to people considering *ex situ* conservation whereby plants were grown outside their habitat, in, for example, a botanic garden or arboretum. As a result, conservation became polarized into two camps, and targets seven and eight of the GSPC reflect this division. Target seven is that 75% of threatened species are conserved *in situ*, while in target eight, 75% of threatened species are conserved *ex situ*, with 10% of these projects in the country of origin of the species concerned.

There is no reason why a declining species should not be conserved in a managed nature reserve. The advantage of this method is that all the other relationships that the plant has with fungi, pollinators, dispersal agents, and other plants are maintained. Furthermore, it is generally believed that more genetic diversity is preserved in *in situ* populations and that the species will therefore be able to evolve as it adapts to changes in conditions. This may all be true, but plants may be far more tolerant of changing pollinators than we previously thought, few plants use a dispersing agent, and the number of plants needed to preserve the genetic diversity of a population of plants is far fewer than you might imagine. Somewhere between 50 and 500 unrelated plants are all you need to preserve 95% of the genetic diversity of a population. The major disadvantage of *in situ* conservation is the security of the site. In the past, this was seen as protection from anthropogenic development or habitat transformation in general. It is now seen that there are two more threats to sites that may be more difficult to control. One is invasion by non-native species, more of which later. The other is climate change.

That the climate is changing cannot be seriously disputed. Whether *Homo sapiens* is behind it and can therefore control it is not important here. The world's plants have experienced climate change before, most recently during the shenanigans at the end of the last ice age. Plants coped presumably by a combination of migration, adaptation, and using hitherto unexploited traits. The relative contributions of these strategies probably varied from species to species and from habitat to habitat. If migration is important, and we know that many plants did migrate long distances during the last three million years, then the encroachment of the urban and agricultural landscapes into natural habitats may have scuppered any chance of migration. The idea of changing the scale of the human landscape at this stage is impractical, but the policy of conserving plants in static nature reserves is equally problematic. Corridors linking them is an idea often proposed, as is landscape conservation whereby the positioning of reserves is coordinated.

All of the advantages of *in situ* conservation can be flipped into the disadvantages of *ex situ* programmes. However, there is a half-way-house known as a species-recovery programme where the goal is to establish a self-propagating population of a species somewhere. In order to achieve this goal, you need to be able to supply everything that the plant requires. You need to understand the habitat requirements, pollination biology, seed storage needs, and you need someone to provide the ongoing, endless monitoring to ensure that the plant survives. This strategy has been championed in Western Australia, where so much cutting-edge conservation happens. Plant species have been brought back from the edge of extinction by species-recovery programmes such as this.

Ex situ conservation has always been regarded as the poor relation to *in situ* conservation, but maybe its time has come in the form of seed banks. While the Millennium Seed Bank Project's (MSBP) first target was the UK flora, it had a secondary target of banking 25,000 species from around the world. That target has been reached and a new more ambitious goal defined. In a world where the future climate is such a big known unknown, seed banks like the MSBP look like a very good idea. All of the criticisms such as genetic erosion, loss of viability, vulnerability to stochastic events like war can be addressed by proper collecting protocols, research, and duplicate collections. It is true to say that much of what we have learnt in the past two decades about seeds is as a result of the rise of seed banks. It may be that in the future seeds stored in seed banks will be used to facilitate assisted migration of plants from

22. Seedlings of the endangered *Encephalartos ferox* being cultivated *ex situ*

one nature reserve to another. It is true that not all seeds appreciate being dried and frozen. Perhaps as many as 30% of species have these so-called recalcitrant seeds, and for these alternatives need to be found.

Seed banks are not a new idea. They began life as gene banks for cultivated varieties of major crop plants, and worldwide there are a number of banks for major crops such as wheat, rice, and maize as well as vegetables. Many of these crops are annuals and so very suited to storage and to the goal of target nine of the GSPC that 70% of the genetic diversity of our crop plants is preserved and protected. Some crops are not propagated by seed, so for crops such as potatoes and fruit trees, field gene banks are the alternative. In many countries, amateur gardeners are keeping the older varieties alive. These may have unique traits that we shall need in the future to impart disease resistance or tolerance of drought.

Mention has already been made of the problems posed by invasive non-native species. This is one of the five major causes of the losses of biological diversity known acronymically as HIPPO: Habitat transformation, Invasive species, Pollution, Population growth, and Over-exploitation. Darwin's theory of evolution through natural selection of organisms that are better suited to their circumstances has often been misrepresented by the sound-bite 'survival of the fittest' when it should be 'survival of the slightly better and the lucky ones'. Darwin observed 150 years ago that nowhere in the world was so well stocked, and the resources so well exploited, that there was no room for another species from abroad. He cited the Cape of South Africa which perhaps has a higher density of species than anywhere else on the planet.

Some people are worried that the fact that the organisms best adapted to grow in South Africa actually evolved in Australia and Europe and vice versa. There are many possible reasons why a non-native plant can outcompete the natives. One is the concept of predator release. This refers to the possibility that plants leave behind the herbivores, pests, and diseases that control them in their native lands and in their new abode they are free from attack. Whatever the reasons, there is a 1 in 1,000 probability that a plant brought to Country B from Country A will become an invasive species and thus cause irreversible damage to the plant communities in Country B. This may not seem short odds, but 70,000 different plants are currently on sale in UK nurseries. These are fully functioning genomes that are potentially far more harmful to other plants than a maize plant that has been genetically modified to resist a herbicide.

Target ten of GSPC 1 was to draw up control measures for the 100 most damaging non-native species. This target has been met, and in GSPC 2 it was modified to be more specific about the control of non-natives in Important Plant Areas in all countries, and more specifically control of further invasions. A problem encountered when controlling non-natives species is predicting which will become invasive. There are reasons why species do not invade; the soil is the wrong pH, the winters too cold, the summers too dry, pollinators fail, dispersal fails, and so on. Unfortunately, though, there is no blueprint for a potential invader; it is only possible to be wise after the event – it is very much like economics in that respect. The only control is prevention. Borders should be sealed to non-native species, as already happens in the USA, Australia, and New Zealand.

The CBD is very clear that we are being greedy when it comes to our consumption of the products of photosynthesis. The Convention on the International Trade in Endangered Species (CITES) should already prevent the exploitation of declining species, but target eleven reiterates that no species of plant should be threatened by trade. Sometimes it is difficult to know which plant is being traded. How can you identify a species when it has been made into a set of window blinds or a herbal remedy? The answer may be close at hand. The practice of DNA barcoding of species has been successful in animals. The idea is that you find a stretch of DNA that varies very little between members of a species, but which varies far more between species. All you have to do then is extract some DNA from the selected region of the genome, sequence it, compare it to the library of this sequence in every known species, and you will either get a name for your sample or be told that you have found a new species. This sounds like science fiction, but it is nearly reality and the prototypes for hand-held machines to do this in the field are being tested. The next step will be to create the library of sequences.

While target eleven addresses the protection of species that are already endangered, the original target twelve aimed at preventing other species being added to the list by trying to ensure that at least 30% of all plant-based products are derived from sustainably harvested suppliers. The revised target twelve is that all wild harvest, plant-based products must be sourced sustainably. Anyone who purchases anything can contribute to the attainment of this target. There are a number of schemes, such as organic food and the FSC (Forestry Stewardship Council) scheme, that enable you to reduce your impact on the world's biological resources. As with many aspects of biology, this area of the GSPC is neither black nor white. The controversy over the production of palm oil in regions of the world inhabited by orang-utans is one such problem. Taking away someone's ability to generate an income is serious action and should not be undertaken lightly.

Hand in hand with the loss of biological resources often goes local, indigenous knowledge. We have a range of phrases that covers this type of information and rarely is it complimentary. Old wives' tales or folklore are just two. It is common to find that this specialist ethnobotanical knowledge is not written down and so is vulnerable when communities are displaced or attracted by the promises of Western knowledge and the Western way of life. Had it not been for the work of ethnobotanists, it may never have been found that the sap of *Strophanthus*, long used as an arrow-tip poison, contained strophanthine that is widely used as a treatment for heart disorders. Target thirteen hopes to stem the tide of ignorance resulting from the careless loss of knowledge, but, with the previous target, this will not be achieved by GSPC 1.

At this point, one comes up against the perennial argument about the interface between poverty and the conservation of biology. Views range from those who think that poverty alleviation is always going to be more important than the existence of a plant species, to those who would be happy to see starvation if that is what the future of plants requires. The reality is, as ever, somewhere in between, and a number of economists are now realizing that economic prosperity is often underpinned by healthy biodiversity. However, to state that poor people are responsible for the destruction of biology and rich people are all caring would be very unfair. The regions of the world where plant species are most threatened can be linked to neither prosperity nor poverty.

23. *Strophanthus*, from which strophanthine is extracted to treat heart disease

The last three targets are too open-ended for us to be sure if a difference has been made, but there are good examples of work that is contributing. Target fourteen urges that the need to conserve our plant heritage is included in education programmes wherever possible. Plants do not have their own voices and so they need an evangelist to speak for them. Botanic gardens, arboreta, and wildlife trusts are just three types of organizations that provide public education programmes that aim to highlight the need to conserve plants and their habitats. Another way to realize the aspirations of the target is to have the conservation of plants written into national curricula for biology courses, and perhaps economics and geography *inter alia*.

The distribution of plants across the world is very uneven and so is the distribution of those people working to conserve these plants on the targets of the GSPC. Target fifteen aims to redress the balance and to ensure that there is a sufficient number of people trained to fulfil the targets and that these people are adequately resourced. It is too commonly stated that no one studies botany any more. If you base this statement on simply the *names* of university courses in the UK, then you may be correct. However, if you take the time to read syllabuses, you will quickly discover that courses in biology, environmental science, conservation, and geography often include all the elements of a 1960s and 1970s botany course. The word 'botany' has a veneer of Victorian parsonages and Edwardian ladies that repels potential students in a way that the newer courses do not. In some countries, however, botany is in the ascendancy. Two such countries are Brazil and Ethiopia. Ten years ago, the project to produce a flora of Ethiopia was staffed entirely by foreigners. Now the work is being carried out entirely by Ethiopians.

It is true to say that projects such as the Millennium Seed Bank have been very effective at creating partnerships and building

24. Training the next generation of field botanists is seen as a priority for many botanic gardens

capacity for plant conservation in many countries. It is vital that the people who are involved in all aspects of plant conservation and the GSPC are working in a coordinated manner. International, regional, national, and local networks are required and are being formed in line with the proposals in the final target, number sixteen. A newly formed network will link people working in the conservation of the plants of oceanic islands. These biological gems often have very high levels of endemism and thus unique floras. They often share common threats. Invasions by non-native species are especially common and damaging on islands. These networks will bring together state organizations, NGOs, international agencies, and local people, as well as amateur botanists and natural historians who are a vital component of many successful national strategies.

No one person or organization is going to solve the problem of declining species on their own. Now that we have the GSPC, we see more clearly where plant conservation has been successful and where the next round of work must be concentrated. It has become a blueprint for other areas of conservation.

We are now in a position to say with confidence that there is no technical reason why a plant species cannot be saved from extinction. In some countries, almost every target of the GSPC will be hit. One such country is the UK, where 400 years of botany and a tradition of field work and botanical recording shows that if there is a sufficient commitment from the indigenous population, then there is a good future for plants and for the humans who will forever depend on these plants for everything.

Further reading

David Beerling, *The Emerald Planet: How Plants Changed Earth's History* (Oxford University Press, 2007)

Eric Chivian and Aaron Bernstein, *Sustaining Life: How Human Health Depends on Biodiversity* (Oxford University Press, 2008)

Charles Darwin *The Origin of Species* (1859)

Jordan Goodman and Vivien Walsh, *The Story of Taxol: Nature and Politics in the Pursuit of an Anti-Cancer Drug* (Cambridge University Press, 2001)

Nicholas Harberd, *Seed to Seed: The Secret Life of Plants* (Bloomsbury, 2007)

Vernon Heywood (ed.), *Flowering Plant Families of the World* (Firefly Books, 2007)

Anna Lewington, *Plants for People* (Eden Project Books, 2003)

D. J. Mabberley, *Mabberley's Plant-Book* (Cambridge University Press, 2008)

Oliver Morton, *Eating the Sun: How Plants Power the Planet* (Fourth Estate, 2007)

Denis J. Murphy, *People, Plants and Genes* (Oxford University Press, 2007)

Michael Proctor, Peter Yeo, and Andrew Lack, *The Natural History of Pollination* (Timber Press, 1996)

Michael G. Simpson, *Plant Systematics* (Elsevier Academic Press, 2006)

Alison M. Smith et al., *Plant Biology* (Garland Science, 2010)

Richard Southwood, *The Story of Life* (Oxford University Press, 2003)

William J. Sutherland and David A. Hill, *Managing Habitats for Conservation* (Cambridge University Press, 1995)

Joel L. Swerdlow, *Nature's Medicine: Plants That Heal* (National Geographic, 2000)

Thomas N. Taylor, Edith L. Taylor, and Michael Krings, *Paleobotany: The Biology and Evolution of Fossil Plants* (Academic Press, 2009)

Plants

Index

Index

Index

ONLINE CATALOGUE

Very Short Introductions

Our online catalogue is designed to make it easy to find your ideal Very Short Introduction. View the entire collection by subject area, watch author videos, read sample chapters, and download reading guides.

THE HISTORY OF LIFE
A Very Short Introduction
Michael J. Benton

There are few stories more remarkable than the evolution of
life on earth. This *Very Short Introduction* presents a succinct
guide to the key episodes in that story - from the very origins
of life four million years ago to the extraordinary diversity of
species around the globe today. Beginning with an explanation
of the controversies surrounding the birth of life itself, each
following chapter tells of a major breakthrough that made new
forms of life possible: including sex and multicellularity, hard
skeletons, and the move to land. Along the way, we witness
the greatest mass extinction, the first forests, the rise of
modern ecosystems, and, most recently, conscious humans.

www.oup.com/vsi

THE EARTH

A Very Short Introduction
Martin Redfern

For generations, the ground beneath the feet of our
ancestors seemed solid and unchanging. Around 30
years ago, two things happened that were to
revolutionize the understanding of our home planet.
First, geologists realized that the continents themselves
were drifting across the surface of the globe and that
oceans were being created and destroyed. Secondly,
pictures of the entire planet were returned from space.
As the astronomer Fred Hoyle had predicted, this
'let loose an idea as powerful as any in history'.
Suddenly, the Earth began to be viewed as a single
entity; a dynamic, interacting whole, controlled by
complex processes we scarcely understood.
It began to seem less solid. As one astronaut put it,
'a blue jewel on black velvet; small, fragile and
touchingly alone'. Geologists at last were able to see
the whole as well as the detail; the wood as well as the
trees. This book brings their account up to date with
the latest understanding of the processes that govern
our planet.

www.oup.com/vsi